NONLINEAR PHYSICAL SCIENCE
非线性物理科学

NONLINEAR PHYSICAL SCIENCE

Nonlinear Physical Science focuses on recent advances of fundamental theories and principles, analytical and symbolic approaches, as well as computational techniques in nonlinear physical science and nonlinear mathematics with engineering applications.

Topics of interest in *Nonlinear Physical Science* include but are not limited to:

- New findings and discoveries in nonlinear physics and mathematics
- Nonlinearity, complexity and mathematical structures in nonlinear physics
- Nonlinear phenomena and observations in nature and engineering
- Computational methods and theories in complex systems
- Lie group analysis, new theories and principles in mathematical modeling
- Stability, bifurcation, chaos and fractals in physical science and engineering
- Discontinuity, synchronization and natural complexity in physical sciences
- Nonlinear chemical and biological physics

SERIES EDITORS

Albert C. J. Luo
Department of Mechanical and Mechatronics Engineering
Southern Illinois University Edwardsville
IL 62026-1805 USA
Email: aluo@siue.edu

Dimitri Volchenkov
Department of Mathematics and Statistics
Texas Tech University
1108 Memorial Circle, Lubbock
TX 79409 USA
Email: dr.volchenkov@gmail.com

INTERNATIONAL ADVISORY BOARD

Edson Denis Leonel

Dynamical Phase Transitions in Chaotic Systems

混沌系统运动状态的切换性

中国教育出版传媒集团

高等教育出版社·北京

Author

Edson Denis Leonel

Departamento de Física

São Paulo State University

Rio Claro, São Paulo, Brazil

**Dynamical Phase
Transitions in
Chaotic Systems**

混沌系统运动状态的
切换性

Hundun Xitong
Yundong Zhuangtai
de Qiehuanxing

图书在版编目（CIP）数据

混沌系统运动状态的切换性 : 英文 / (巴西) 埃德
森 · 丹尼斯 · 利奥 (Edson Denis Leonel) 著 . — 北京 :
高等教育出版社 , 2024.1
（非线性物理科学）
ISBN 978-7-04-061663-7

Ⅰ . ①混… Ⅱ . ①埃… Ⅲ . ①混沌理论—英文 Ⅳ .
① O415.5

中国国家版本馆 CIP 数据核字 (2024) 第 004235 号

策划编辑 李 鹏　　责任编辑 李 鹏　　封面设计 杨立新
责任印制 田 甜

出版发行 高等教育出版社
社址 北京市西城区德外大街 4 号
邮政编码 100120
购书热线 010-58581118
咨询电话 400-810-0598
网址 http://www.hep.edu.cn
　　　http://www.hep.com.cn
网上订购 http://www.hepmall.com.cn
　　　http://www.hepmall.com
　　　http://www.hepmall.cn
印刷 涿州市京南印刷厂

开本 787mm×1092mm 1/16
印张 6.25
字数 130 千字
版次 2024 年 1 月第 1 版
印次 2024 年 1 月第 1 次印刷
定价 99.00 元

To my son Gustavo

Preface

The main goal of this book is to discuss some scaling properties and characterize two-phase transitions in nonlinear systems described by mappings. The chaotic dynamics is given by the unpredictability of the time evolution of two very close initial conditions in the phase space. It yields an exponential divergence from each other as time passes. The chaotic diffusion is investigated in a specific family of area-preserving maps. It is also considered for two dissipative systems, leading to a scaling invariance, which is proved to be a characteristic of a phase transition. Two different types of transitions are considered in the book. One gives a transition from integrability to non-integrability and is observed in a two-dimensional, nonlinear, and area-preserving mapping, hence conservative dynamics. The dynamical variables considered are the action and angle. A sine function dictates the nonlinearity of the mapping. The angle exhibits a property that it diverges continuously in the limit when the action approaches zero. As will be shown, this characteristic leads to the existence of a chaotic sea, producing chaotic diffusion in the phase space. A property of the phase space is that the chaotic sea is limited by a set of invariant spanning curves, hence limiting the chaotic diffusion.

Another transition considers the dynamics given by nonlinear mappings and describes the suppression of the unlimited chaotic diffusion for a dissipative standard mapping. The dynamics is characterized by a pair of dynamical variables, action, and angle. A sine function gives the nonlinearity of the dynamics and is controlled by a control parameter providing the step length for the amplitude of the motion. The determinant of the Jacobian matrix is smaller than one creating attractors in the phase space in the presence of dissipation, leading to the suppression of the diffusion.

The suppression of unlimited diffusion is also observed in a far more complicated system describing the dynamics of a particle in a billiard with a moving boundary. In a billiard system, a particle or, in an equivalent way, an ensemble of noninteracting particles moves inside a closed region with which a particle collides. When the collisions are elastic, energy and momentum are preserved in the moving referential frame. Depending on the initial condition as well as on the control parameters, the particle may exhibit an unbounded growth of energy that is called Fermi acceleration. On the other hand, fractional energy loss happens when inelastic collisions are taken into account, and only momentum is conserved in the moving referential frame at the impact. The phenomenon of unlimited energy growth is not robust since tiny dissipation is enough to suppress the unbounded growth, leading to a transition from limited to unlimited energy growth.

During the investigation of the transitions, three different procedures are assumed, corroborating for a scaling invariance and hence, for a phase transition. The first procedure takes the characterization by using scaling hypotheses described by a homogeneous and generalized function leading to a set of three critical exponents related to each other via a scaling law. The second procedure uses a semi-phenomenological scheme estimating the position of the first invariant spanning curve for a conservative mapping. The design is connected with a transition well known in the non-dissipative standard mapping, a transition from locally chaotic to globally chaotic dynamics. This estimation allows for determining the size of the chaotic sea and obtaining one of the three critical exponents. A transformation of the equation of differences in an ordinary differential equation allowing direct integration furnishes the slope of acceleration, therefore another critical exponent. The third one can be obtained by using a scaling law determined by the homogeneous and generalized function.

The last formalism used the solution of the diffusion equation to obtain the probability density of observing a given action at a specific time. From the expression of the probability density, all momenta are determined as a function of the parameters as well as the time.

I wrote this book considering the discussion and characterization of the two mentioned transitions. The first one is a transition from integrability to non-integrability. It was investigated in a conservative mapping. A second one is a transition from limited to unlimited diffusion, which appeared to be classified after my book *Scaling Laws in Dynamical Systems*, was published by Higher Education Press and Springer in 2021, which served as the primary textbook basis for these notes. The different papers published by my group were also used to guide the discussions along with the chapters. The book was edited to be original enough to contribute to the existing literature but with no excessive superposition of the topics already considered in different other textbooks. The organization of the book is as follows.

In Chap. 1, we pose the three problems discussed in the book and the formalism considered in each characterization.

Chapter 2 is devoted to constructing a nonlinear mapping from two-degree of freedom Hamiltonian. Applications involving specific functions will be presented.

In Chap. 3, we present a phenomenological discussion that is used to determine a scaling law. It is based on three scaling hypotheses allied to a homogeneous and generalized function that depends on three characteristic exponents, which are applied to an area-preserving mapping.

Chapter 4 discussed a semi-phenomenological procedure to estimate the localization of the first invariant spanning curve. The position of the curve defines the length of the chaotic sea, hence a limitation for the chaotic diffusion. Since it is obtained as a function of the control parameters, one of the three critical exponents is found analytically. An approximation that transforms the equation of the mapping, written in terms of an equation of differences, into a differential equation turns easy to determine the acceleration exponent, consequently one of the critical exponents obtained by the solution of an ordinary differential equation.

In Chap. 5, we regard the diffusion equation and obtain a solution that describes the chaotic diffusion in a chaotic sea for the conservative mapping. It is given by imposing two boundary conditions and a specific initial condition. The probability density's momenta of the distribution are obtained from the knowledge of the probability density. The scaling properties are all described together with critical exponents and a scaling law.

Chapter 6 discusses the phase transition from integrability to non-integrability in an area-preserving mapping. Four main questions are answered: (1) Identify the broken symmetry; (2) Define the order parameter; (3) Discuss the elementary excitation; and (4) Discuss the topological defects which impact the transport of particles.

In Chap. 7, we discuss a dissipative standard mapping with the initial goal of characterizing the behavior of the chaotic diffusion for particles moving along a chaotic attractor created by the introduction of dissipation in one of the equations of the mapping. The scaling hypotheses are discussed, and two scaling laws are derived via an association with a homogeneous and generalized function. An analytical formalism is also discussed, leading to acquiring the critical exponents. The solution of the diffusion equation is also obtained by imposing boundary and initial conditions.

Chapter 8 is reserved for discussing the phase transition properly. An order parameter is obtained as well as its susceptibility. The elementary excitation is determined while a discussion on the topological defects that impact particles' transport is made. It also presents a dialogue concerning the break of symmetry at the transition.

Chapter 9 presents an elementary discussion on the time-dependent billiard. The equations of the mapping are determined, and a discussion on the unlimited energy growth is presented.

Chapter 10 is devoted to introducing inelastic collisions of the particle with the boundary, constructing the mapping equations by using momentum conservation law, and presenting the scaling hypotheses together with the determination of the critical exponents. A scaling invariance is also described.

In Chap. 11, the discussion on the phase transition from limited to unlimited energy growth is made for a dissipative time-dependent billiard.

I typed all the notes from the title until the last word of the references using LATEX. As graphical editors, I used *xmgrace* and *gimp*, in almost all figures.

Rio Claro, São Paulo, Brazil   Edson Denis Leonel
December 2022

Acknowledgements

The primary motivation to prepare these notes came from investigating some statistical properties for chaotic dynamics in two-dimensional and nonlinear mappings when a surprising scaling invariance for the diffusion emerged. Then we used different procedures to describe the phenomena, including a phenomenological approach using a set of scaling hypotheses leading to a scaling law relating three critical exponents. A semi-phenomenological procedure led to the localization of invariant spanning curves in the phase space. They estimate the size of the chaotic sea in the conservative case, implying a finite diffusion and allowing direct estimation of a critical exponent. In this procedure, one of the equations of differences is transformed into a differential equation, which analytically gives another critical exponent. The use of a scaling law can determine the third critical exponent. Lastly, the solution of the diffusion equation gives the probability density of observing a particle with a given action at a given time. All the momenta can be obtained from the probability density, and all the scaling properties are confirmed.

All the above results describe the behavior of the scaling invariance present in the diffusion observed for chaotic particles in nonlinear and area-preserving mapping. Scaling invariance has also been observed in dissipative systems, particularly in a transition from limited to unlimited chaotic diffusion. Interestingly, the scaling invariance is a characteristic of a phase transition. Therefore, the book's primary goal is to discuss two types of phase transition present in nonlinear mappings: (i) transition from integrability to non-integrability; and (ii) transition from limited to unlimited diffusion. The first is made in a conservative system, while the second uses two dissipative systems with different complexities.

I acknowledge my long-term collaborators for constant discussions including Professors Makoto Yoshida, Juliano Antonio de Oliveira, José Antonio Méndez-Bermúdez, Denis Gouvêa Ladeira, Iberê Luiz Caldas, and Mário Roberto da Silva, my former Ph.D. and Master students Diego Fregolent Mendes de Oliveira, André Luiz Prando Livorati, Diogo Ricardo da Costa, Matheus Hansen Francisco, Célia Mayumi Kuwana, Joelson Dayvison Veloso Hermes, Yoná Hirakawa Huggler and my present students Felipe Augusto Oliveira Silveira, Anne Kétri Pasquinelli da Fonseca, Lucas Kenji Arima Miranda.

I kindly acknowledge the Department of Physics of Unesp in Rio Claro for providing the conditions for the present material's construction and edition.

Contents

List of Figures

Chapter 1
Posing the Problems

Abstract This chapter aims to introduce the problems addressed in the book.

1.1 Initial Concepts

Scaling features are observed and applied in a variety of fields and several areas of research. One of the outstanding characteristics of continuous phase transitions is the so-called scaling invariance. It implies that a given phenomenon turns out to be independent of the scale with which it is observed, mostly while seen concerning critical control parameters. In a continuous phase transition, a qualitative change in the behavior of a given observable happens at a point where the properties are singular but continuous.

Giving a scaling invariance is observed when the dynamics of a given system passes a phase transition, the main focus of the present investigation is to characterize two specific phase transitions: (i) a transition from integrability to non-integrability and; (ii) a transition from limited to unlimited diffusion of a dynamical variable. Both transitions are observed in the dynamics described by nonlinear mappings. The first is noticed in a conservative system, while the other is investigated when the system passes from conservative to non-conservative dynamics. The scaling invariance is observed and characterized via chaotic diffusion.

Case (i) considers the dynamics given by a Hamiltonian $H = H_0 + \epsilon H_1$, where H_0 gives the integrable part, while H_1 corresponds to the non-integrable part and ϵ is the control parameter. For the case of $\epsilon = 0$, the system is integrable while it is non-integrable for any $\epsilon \neq 0$. Because the energy is a constant of motion, the flux of solutions lies in a $3D$ surface. When it is intercepted by a plane of constant angle, the dynamics is furnished by using a two-dimensional and nonlinear mapping in the angle-action variables. A careful choice of a family of area-preserving maps leads the angle to diverge in the limit of vanishingly action. This mapping property guarantees the existence of a chaotic sea in the phase space, which births for any $\epsilon \neq 0$ as mixed and contains periodic islands surrounded by a chaotic sea confined by a set of invariant spanning curves. The lowest invariant spanning curve is crucial in limiting the size of the chaotic diffusion and leads to an essential scaling property. The chaotic diffusion of particles moving along the chaotic sea can diffuse up to a specific limit.

The diffusion saturation obeys a power law in the control parameter, with the slope defining a critical exponent. The saturation value for the chaotic diffusion represents an important observable called the order parameter. In a continuous phase transition, such parameter goes continuously to zero. In distinction, its susceptibility, which furnishes the response of the order parameter to a variation of the control parameter, diverges at the phase transition. Moreover, the intensity of the nonlinearity gives the elementary excitation of the dynamics.

In case (ii), the dynamics is described by the standard mapping, also called the Chirikov–Taylor map, which is reported in terms of a two-dimensional mapping provided by a pair action I and angle θ. Depending on the control parameter ϵ, invariant spanning curves are not present in the dynamics, allowing unlimited diffusion of the action to be observed. This phenomenon is suppressed by the introduction of small dissipation yielding in a violation of Liouville's theorem, creating attractors in the phase space. Since such attractors are far from infinity, a limited diffusion is observed, characterizing a transition from limited to unlimited diffusion.

Due to the properties of the chaotic attractor, the chaotic diffusion reaches a saturation value that depends on the amount of dissipation. The saturation is given by a power law of the dissipation strength. The exponent characterizing the saturation behavior defines a critical exponent, bringing relevant information concerning the phase transition.

Yet in case (ii), a phenomenon that can also be described via suppression of unlimited diffusion happens to a class of dynamical systems classified as billiards with time-dependent boundaries. According to the Loskutov–Ryabov–Akhinshin (LRA) conjecture, a chaotic dynamics of billiard with a static border is a sufficient condition to observe Fermi acceleration when a time-dependent perturbation to the boundary is added. Indeed, Fermi acceleration is a phenomenon in which a particle, or an ensemble of non-interacting particles, acquires unlimited energy after many collisions with a time-dependent wall. The endless energy growth is marked by a regime of unbounded growth of the squared velocity, hence unlimited diffusion. Four-dimensional mappings generally describe billiard systems that exhibit such characteristics with two pairs of dynamical variables. One pair characterizes the angular variables, while the other is the velocity and time.

The unlimited diffusion, however, is not robust since introducing an infinitesimal dissipation such as inelastic collision is enough to suppress the unbounded energy growth. The suppression leads to the creation of attractors, hence producing limited diffusion for chaotic dynamics. The scaling properties are extracted for the dynamics along the chaotic attractor and show dynamics enlightening qualitative behavior similar to that observed in the transition from limited to unlimited diffusion for two-dimensional mappings; however, they are observed for a rather complicated system.

Near both transitions, our aim to investigate and characterize a phase transition lies in answering four main questions: (1) What is the break of symmetry involved in the transition? (2) What is a possible order parameter and its susceptibility (that is, the answer of the order parameter due to the variation of an external field—or control parameter)? (3) What are the elementary excitations of the systems letting the particles diffuse? (4) What are the topological defects impacting the transport of particles?

Even though the terms mentioned above may differ in the areas, they are intuitive in the investigation. For the conservative mapping, two indications of the symmetry break are identified, one in the phase space and the other in the dynamical equation. The integrable regime's phase space is foliated while mixed for the non-integrable case. The chaotic dynamics is limited to a set of invariant spanning curves whose first one has a crucial importance in the scaling property noticed in the diffusion. The second one is observed in the equation describing the dynamical variable associated with the action. It also passes from a symmetry break with a nonnull control parameter where the left-hand side of the equation differs from the right side. The dissipation limits the dynamical variable to reach a finite range for the transition from limited to unlimited diffusion. The curve of unlimited growth then experiences a crossover that limits the unbounded growth, destroying its growing symmetry.

The terminology defining an order parameter comes from statistical mechanics, particularly magnetic materials, mainly composed of individual spins occupying fixed positions that are allowed to orientate due to external (or even from local configurational) fields. Each spin may align to an external field, making the system ordered. This configuration produces a nonnull observable called magnetization. However, as soon as the temperature increases, it gives energy to the spins allowing them to fluctuate their magnetic orientation around their average position. There is a critical temperature where no resulting alignment is observed in the material producing a disordered phase. At this configuration, the material's magnetization is zero, and it goes smoothly to such an estate. Magnetization is the natural variable describing ordered phases distinct from disordered ones. The magnetization's response to the external field variation defines its susceptibility. Nevertheless, it diverges in the limit as the magnetization goes to zero, which marks one of the characteristics of a continuous phase transition. In the problems discussed in the book, the transitions are linked mainly to diffusion. The observables leading to the order parameter in the two transitions are related to the stationary estate of the diffusion.

In a random walk process, a particle (walker) is suspended in a fluid and moves randomly around its position with a limited step size ℓ. For one-dimensional motion, there is a probability p a particle moves to the right and a probability q it moves to the left. Of course $p + q = 1$. When $p \neq q$, the motion has a bias, and a preferred direction dominates over the other. The particle moves due to random processes that might be collisions of the particle under suspension with particles or molecules of the fluid. The limited length of each step turns out to be, in this case, the elementary excitation producing the particle's motion, leading to diffusion. In the problems treated in the book, diffusion is produced by specific terms present in the dynamical equations. Most defining the intensity of the nonlinearity is related to the elementary excitations in the transitions investigated.

In statistical mechanics, topological defects may be numerous. They may range from a simple additional column of atoms present in a piece of material causing a dislocation; they are observed in magnetic material where a magnetization from a specific point points out in all directions. They are observed too in nematic crystals and many other materials and systems. The problems discussed in the book are mostly related to regions of the phase space where the different ways of the transport of particles affect diffusion. In a conservative system where the phase space has mixed form, chaos, periodic islands, and invariant tori are all observed. Each one of them is obtained by evolving different initial conditions. However, a dynamical regime called stickiness directly affects the transport of particles. Stickiness is observed when a particle moving along a chaotic domain passes close enough to an island or an invariant curve and stays moving around, hence temporarily confined. The confinement may be short or not, in case the diffusion is affected. For dissipative dynamics where no islands are present, the diffusion may be affected by periodic attractors corresponding to a set of points where trajectories converge sufficiently long.

1.2   Summary

In summary, in this chapter, we presented the outlines that will be discussed throughout the book.

Chapter 2
A Hamiltonian and a Mapping

Abstract In this chapter, we discuss some steps to obtain a nonlinear and area-preserving mapping by considering a two-degree of freedom Hamiltonian. Specific choices made with the functions lead to describing different systems. Imposing the determinant of the Jacobian matrix equal to unity leads to area preservation. Chaotic diffusion is considered via different techniques leading to a scaling invariance.

2.1 The Mapping

We consider some dynamical properties of the phase space in a family of mappings obtained considering a two-degree of freedom Hamiltonian. In the characterization, the Hamiltonian is comprised of two parts. One of them is assumed to be integrable. At the same time, the other one is linked to non-integrability and is described by a control parameter ϵ. It controls a transition from integrability to non-integrability. We consider a Hamiltonian of the type

$$H(I_1, I_2, \theta_1, \theta_2) = H_0(I_1, I_2) + \epsilon H_1(I_1, I_2, \theta_1, \theta_2), \tag{2.1}$$

where I_i and θ_i, $i = 1, 2$ are conjugated variables. H_0 gives the integrable part, while H_1 furnishes the nonintegrable part, controlled by a parameter ϵ. One notices that for $\epsilon = 0$, the system is integrable. It is non-integrable for $\epsilon \neq 0$. The control parameter ϵ defines a transition between two regimes: (i) integrability for $\epsilon = 0$; (ii) non-integrability when $\epsilon \neq 0$.

The integrable case happens for $\epsilon = 0$. The Hamiltonian assumes the form $H(I_1, I_2, \theta_1, \theta_2) \rightarrow H(I_1, I_2) = H_0(I_1, I_2)$, which is independent of θ_1 and θ_2. In this case, the Hamiltonian equations are

$$\frac{dI_1}{dt} = \dot{I}_1 = -\frac{\partial H_0}{\partial \theta_1} = 0,$$
$$\frac{dI_2}{dt} = \dot{I}_2 = -\frac{\partial H_0}{\partial \theta_2} = 0, \tag{2.2}$$

and the solutions for $I_i = I_i(\theta_i)$ with $i = 1, 2$ are constant. The other two equations are written as

$$\frac{d\theta_1}{dt} = \dot{\theta}_1 = -\frac{\partial H_0}{\partial I_1} = f_1(I_1, I_2),$$

$$\frac{d\theta_2}{dt} = \dot{\theta}_2 = -\frac{\partial H_0}{\partial I_2} = f_2(I_1, I_2), \tag{2.3}$$

where $f_i(I_1, I_2)$, with $i = 1, 2$ are assumed to be functions of I_1 and I_2 and are time-independent.

The case $\epsilon \neq 0$ leads to non-integrability. Considering that Eq. (2.1) is independent of time, the energy is a constant of motion. The system is described by four dynamical variables I_1, I_2, θ_1, and θ_2 and assuming the energy as a constant, one can reduce the 4D flux of solutions to a relevant set of three variables, indeed I_1, θ_1, and θ_2. Intercepting the flux of solutions by a plane I_1 versus θ_1 called Poincaré section obtained by considering θ_2 constant, the set of points intercepting the plane can be described by a discrete mapping.

A generic two-dimensional mapping describing the dynamics of the Hamiltonian given by Eq. (2.1) is written as

$$\begin{cases} I_{n+1} = I_n + \epsilon h(\theta_n, I_{n+1}), \\ \theta_{n+1} = [\theta_n + K(I_{n+1}) + \epsilon p(\theta_n, I_{n+1})] \mod(2\pi), \end{cases} \tag{2.4}$$

where h, K, and p are the nonlinear functions of their variables whose index n identifies the nth iteration of the mapping. Considering the mapping (2.4) originated from a Hamiltonian, the phase space's area has to be preserved. Consequently, functions $h(\theta_n, I_{n+1})$ and $p(\theta_n, I_{n+1})$ must follow a relation between themselves, which is obtained via the determinant of the Jacobian matrix equals to ± 1. The elements of the matrix are written as

$$\frac{\partial I_{n+1}}{\partial I_n} = \frac{1}{1 - \epsilon \frac{\partial h(\theta_n, I_{n+1})}{\partial I_{n+1}}}, \tag{2.5}$$

$$\frac{\partial I_{n+1}}{\partial \theta_n} = \epsilon \frac{\partial h(\theta_n, I_{n+1})}{\partial \theta_n} + \epsilon \frac{\partial h(\theta_n, I_{n+1})}{\partial I_{n+1}} \frac{\partial I_{n+1}}{\partial \theta_n}, \tag{2.6}$$

$$\frac{\partial \theta_{n+1}}{\partial I_n} = \left[\frac{\partial K(I_{n+1})}{\partial I_{n+1}} + \epsilon \frac{\partial p(\theta_n, I_{n+1})}{\partial I_{n+1}} \right] \frac{\partial I_{n+1}}{\partial I_n}, \tag{2.7}$$

$$\frac{\partial \theta_{n+1}}{\partial \theta_n} = 1 + \epsilon \frac{\partial p(\theta_n, I_{n+1})}{\partial \theta_n}$$

$$+ \left[\frac{\partial K(I_{n+1})}{\partial I_{n+1}} + \epsilon \frac{\partial p(\theta_n, I_{n+1})}{\partial I_{n+1}} \right] \frac{\partial I_{n+1}}{\partial \theta_n}. \tag{2.8}$$

From the elements of the Jacobian matrix, the determinant of the matrix is given by

$$\det J = \frac{\left[1 + \epsilon \frac{\partial p(\theta_n, I_{n+1})}{\partial \theta_n} \right]}{\left[1 - \epsilon \frac{\partial h(\theta_n, I_{n+1})}{\partial I_{n+1}} \right]}. \tag{2.9}$$

Accepting the determinant is equal to 1, then

$$\frac{\partial p(\theta_n, I_{n+1})}{\partial \theta_n} + \frac{\partial h(\theta_n, I_{n+1})}{\partial I_{n+1}} = 0. \tag{2.10}$$

In the next section, we describe some applications of specific choices of functions p, h, and K.

2.2 Specific Applications

We presume as fixed the terms $p(\theta_n, I_{n+1}) = 0$ and $h(\theta_n) = \sin(\theta_n)$. Different applications depend on the expressions for the function K and will be made in sequence.

2.2.1 Standard Mapping

The expression of K for the standard mapping, also called Chirikov–Taylor mapping, is given by $K(I_{n+1}) = I_{n+1}$ that leads to the following mapping:

$$T_{sm} : \begin{cases} I_{n+1} = I_n + \epsilon \sin(\theta_n), \\ \theta_{n+1} = [\theta_n + I_{n+1}] \mod(2\pi). \end{cases} \tag{2.11}$$

An orbit is a chronological sequence of an initial condition (I_0, θ_0) after many applications of the operator T such that given a pair of $(I_0, \theta_0) \rightarrow (I_1, \theta_1) \rightarrow (I_2, \theta_2) \ldots \rightarrow (I_n, \theta_n) \ldots$. A set of all combinations of orbits leads to the phase space showing all possible estates for the dynamics. Figure 2.1 shows four phase space plots assuming different control parameters, as labeled in the figure.

Figure 2.1a was constructed for $\epsilon = 0$. It shows that the curves are parallel due to the integrability. The dynamical variable I is a constant, together with the energy, as was the premise used in the mapping construction. Figure 2.1b was plotted for the control parameter $\epsilon = 0.1$ and the integrability is no longer observed.

The phase space reveals an elliptic fixed point surrounded by closed curves and a set of invariant spanning curves crossing the entire domain. For the definition of the figure, a chaotic sea is not observed in the scale of the figure. Increasing the control parameter to $\epsilon = 0.5$, as shown in Fig. 2.1c, other periodic orbits emerge, as is the case of a chain of period-two islands. For the control parameter $\epsilon = 0.97$, chaos is now observed at the scale of the figure, as shown in Fig. 2.1d.

There are two critical transitions observed in the standard mapping. One of them occurs when ϵ alters from $\epsilon = 0$ to $\epsilon \neq 0$. It then characterizes a transition from integrability to non-integrability. Another transition happens when the control parameter achieves a critical one $\epsilon_c = 0.9716\ldots$. For parameters smaller than $\epsilon < \epsilon_c$, the

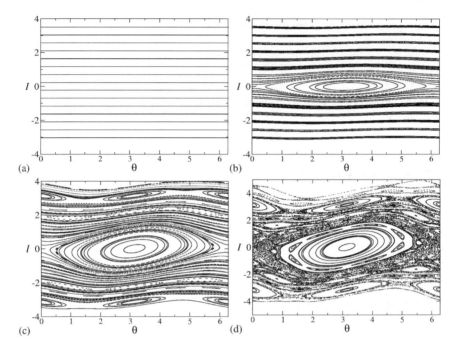

Fig. 2.1 Plot of the phase space for the mapping (2.11) considering the control parameters: **a** $\epsilon = 0$; **b** $\epsilon = 0.1$; **c** $\epsilon = 0.5$ and **d** $\epsilon = 0.97$

phase space exhibits mixed form such that chaotic seas surround periodic islands, and invariant spanning curves exist. For $\epsilon > \epsilon_c$, the invariant spanning curves no longer exist, and the chaotic sea may diffuse unbounded in the phase space. One notices a transition from local chaos for $\epsilon < \epsilon_c$ to global chaos for $\epsilon > \epsilon_c$. Depending on the initial condition, unbounded diffusion for the dynamical variable I can be observed in the latter condition.

The fixed points are obtained setting that $I_{n+1} = I_n = I$ and $\theta_{n+1} = \theta_n = \theta + 2m\pi$, leading then to two sets of fixed points: (1) $(I, \theta) = (2m\pi, 0)$ with $m = \pm 1, \pm 2, \pm 3, \ldots$ which are unstable; (2) $(I, \theta) = (2m\pi, \pi)$ with $m = \pm 1, \pm 2, \pm 3, \ldots$ which are elliptical, consequently stable for $0 \leq \epsilon < 4$ and unstable for any $\epsilon > 4$. As predicted, the determinant of the Jacobian matrix is the unity leading to area preservation of the phase space.

2.2.2 The Fermi–Ulam Model

The Fermi–Ulam model was proposed as a mathematical toy model in an attempt to conceivably describe the acceleration of cosmic particles by the moving magnetic fields of the cosmos. By using the Hamiltonian formalism, the function K is written as $K(I_{n+1}) = 2/I_{n+1}$ guiding to the following expression of the mapping[1]

$$T_{fm} : \begin{cases} I_{n+1} = I_n + \epsilon \sin(\theta_n), \\ \theta_{n+1} = [\theta_n + \frac{2}{I_{n+1}}] \mod(2\pi). \end{cases} \tag{2.13}$$

A plot of the phase space is shown in Fig. 2.2 for the parameters: (a) $\epsilon = 10^{-2}$ and (b) $\epsilon = 10^{-3}$. We notice that the phase is mixed and contains a set of periodic islands surrounded by a chaotic sea confined by a set of invariant spanning curves. Since the determinant of the Jacobian matrix is unity, area preservation is observed in the phase space.

The fixed points for mapping (2.13) have the following expressions: (1) $(I, \theta) = (\frac{1}{m\pi}, 0)$ which are elliptical, hence stable and; (2) $(I, \theta) = (\frac{1}{m\pi}, \pi)$, which are always unstable.

The analytical expression of the function K implies an intriguing physical situation in the phase. Since the dynamical variable θ is modulated 2π, for small values of I yields the function $K = 2/I$ to assume large values thus turning uncorrelated the variables θ_{n+1} and θ_n. Since θ is the argument of the sine function, one notices a diffusion in the variable I. Nevertheless, as soon as I increases, the ratio $2/I$ becomes small and brings a correlation between θ_{n+1} and θ_n, producing regularity in the phase space, including the existence of periodic islands and invariant spanning curves. These invariant spanning curves, particularly the lowest energy one, imply a crucial scaling present in the chaotic diffusion for the particles in the phase space. Their presence limits the chaotic sea, therefore imposing a saturation for the diffusion.

2.2.3 The Bouncer Model

An alternative model to investigate Fermi acceleration considers the reinjection of the particle for an additional collision with the moving wall, which considers a

---

[1] In the original formulation, the mapping is assembled in the variables velocity V and phase ϕ, hence an application $T(V_n, \phi_n) = (V_{n+1}, \phi_{n+1})$ whose expression is given by

$$\begin{cases} V_{n+1} = |V_n| + \epsilon' \sin(\phi_{n+1}), \\ \phi_{n+1} = [\phi_n + \frac{2}{V_n}] \mod(2\pi). \end{cases} \tag{2.12}$$

The absolute value introduced in the variable velocity considers the dynamics of the static wall approximation and $\epsilon' = 2\epsilon$. Indeed, the system is composed of a particle or an ensemble of non-interacting particles restricted to moving inside two walls. One is fixed, and the other one exchanges energy with the particle. While the amplitude of the moving wall is small, the static wall approximation speeds up the numerical simulation by assuming both walls are fixed and one of them, say the one on the left, can swap energy as if it was moving. However, a non-physical occurrence must be avoided. After an impact, when the particle was permitted to have negative velocities, a situation attainable to be observed in the full model, the one that considers the natural movement of the wall, is not allowed to happen in the static wall approximation. The absolute value is introduced in the first equation of the mapping to avoid this occurrence.

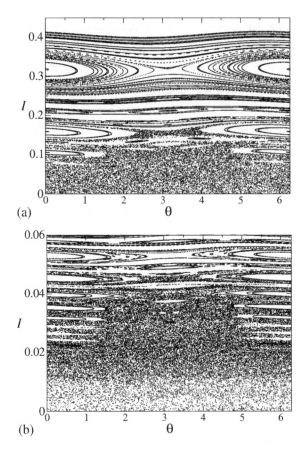

(a)

(b)

Fig. 2.2 Plot of the phase space for the mapping (2.13) considering the control parameters: **a** $\epsilon = 10^{-2}$; **b** $\epsilon = 10^{-3}$

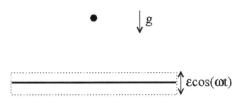

Fig. 2.3 Sketch of the bouncer model. The amplitude of motion is ϵ which moves with a frequency ω, while g identifies a constant gravitational field

constant gravitation field. Figure 2.3 shows a sketch of the bouncer model. Assuming a static wall approximation, the one which considers the amplitude of the motion is small at the point the wall can be thought of as fixed but, at the instant of the impact, it transfers energy and momentum as if it were moving, leading to assume in

Hamiltonian formalism $K(I_{n+1}) = \zeta I_{n+1}$, assuming ζ is a constant, with a mapping given by

$$T_b : \begin{cases} I_{n+1} = I_n + \epsilon \sin(\theta_n), \\ \theta_{n+1} = [\theta_n + \zeta I_{n+1}] \mod(2\pi). \end{cases} \tag{2.14}$$

The phase space has similar properties to those observed in the standard mapping, including a transition from limited to unlimited energy growth, leading to Fermi acceleration.

2.2.4 A Hybrid Fermi–Ulam Bouncer Model

A hybrid version of the Fermi–Ulam and bouncer model consists of a composition of the two models, which contemplates a classical particle moving confined between two walls under the action of a constant gravitational field. The walls are parallel to each other, where one is located at the origin, oscillating around an average value. The other is located in the vertical from a distance ℓ. For low values of the particle's velocity, the influence of the bouncer model is more considerable than the Fermi–Ulam when the speed is large. Two reinjection mechanisms are present in the model. One of them is the gravitational field which always acts, and the other is a collision of the particle with the fixed wall, which happens only when the velocity of the particle allows it to. For large velocities, the stationary wall reinjects the particle for a further impact with the wall transferring energy to the particle.

When we consider the Hamiltonian formalism, the function K is written as

$$K(I_{n+1}) = \begin{cases} 4(\pi N_c)^2 (I_{n+1} - \sqrt{I_{n+1}^2 - \frac{1}{(\pi N_c)^2}}) \text{ if } I_{n+1} > \frac{1}{(\pi N_c)}, \\ 4(\pi N_c)^2 I_{n+1} \text{ if } I_{n+1} \leq \frac{1}{(\pi N_c)}, \end{cases}$$

with (πN_c) constant.

The mapping[2] is then written as

$$T_{hfub} : \begin{cases} I_{n+1} = I_n + \epsilon \sin(\theta_n), \\ \theta_{n+1} = [\theta_n + K(I_{n+1})] \mod(2\pi). \end{cases} \tag{2.15}$$

The fixed points for mapping (2.15) are illustrated in Table 2.1.

---

[2] In the original version, the dynamical variables are velocity V and phase ϕ. Because the easy form of $K(I_{n+1})$ considers only the static wall approximation, in the original version, there is also an absolute function in the velocity to avoid, similar to what is observed in the Fermi–Ulam and bouncer model, non-physical cases of negative velocity after the collision.

Table 2.1 Table of fixed points for the hybrid Fermi–Ulam bouncer model

Period	I	ϕ	ϵ	Type
1	$\frac{1}{j\pi}$, $j = 1, 2, 3, \dots$	0	All	Hyperbolic
1	$\frac{1}{j\pi}$, $j = 1, 2, 3, \dots$	π	$< \frac{1}{j^2\pi^2}$	Elliptic
1	$\frac{1}{j\pi}$, $j = 1, 2, 3, \dots$	π	$= \frac{1}{j^2\pi^2}$	Parabolic
1	$\frac{1}{j\pi}$, $j = 1, 2, 3, \dots$	π	$> \frac{1}{j^2\pi^2}$	Hyperbolic

2.2.5 Family of Mappings

Let us now discuss the case of $h(\theta_n, I_{n+1}) = \sin(\theta_n)$ and $K = 1/|I_{n+1}|^\gamma$ with $\gamma > 0$ and $p(\theta_n, I_{n+1}) = 0$, which will be considered the mapping for describing the transition from integrability to non-integrability. The equations of the mapping are

$$\begin{cases} I_{n+1} = I_n + \epsilon \sin(\theta_n), \\ \theta_{n+1} = [\theta_n + \frac{1}{|I_{n+1}|^\gamma}] \mod(2\pi), \end{cases} \tag{2.16}$$

where $\epsilon > 0$ is a control parameter.

Figure 2.4 illustrates the phase space for the mapping (2.16) using $\epsilon = 0.01$ and $\gamma = 1$. The phase space shows stability islands surrounded by a chaotic region limited by invariant spanning curves. Area preservation does not allow a particle moving along the chaotic sea to intrude on any island nor cross the invariant spanning curves, otherwise would lead to a violation of Liouville's theorem.

Fig. 2.4 Plot of the phase space for the mapping (2.16) considering the control parameters $\epsilon = 0.01$ and $\gamma = 1$. The symbols identify the elliptic fixed points

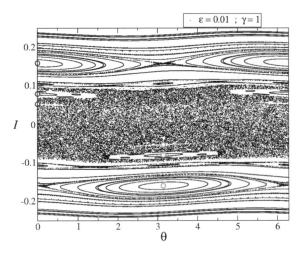

The fixed points are

$$(\theta, I) = \begin{cases} 0, \pm \left(\frac{1}{2m\pi}\right)^{\frac{1}{\gamma}}, \\ \pi, \pm \left(\frac{1}{2m\pi}\right)^{\frac{1}{\gamma}}. \end{cases} \tag{2.17}$$

An elliptic fixed point is given for $(0, (1/(2m\pi))^{1/\gamma})$ and $(\pi, -(1/(2m\pi))^{1/\gamma})$ when

$$m < \frac{1}{2\pi} \left[\frac{4}{\epsilon\gamma} \right]^{\frac{\gamma}{\gamma+1}}. \tag{2.18}$$

The circles in Fig. 2.4 give the elliptic fixed points for $m = 1, 2, 3$. For

$$m \geq \frac{1}{2\pi} \left[\frac{4}{\epsilon\gamma} \right]^{\frac{\gamma}{\gamma+1}}, \tag{2.19}$$

the fixed points $(0, -(1/(2m\pi))^{1/\gamma})$ and $(\pi, (1/(2m\pi))^{1/\gamma})$ are hyperbolic, therefore unstable.

2.3 Summary

In this chapter, the expressions of two-dimensional, area-preserving mappings were obtained using a two-degree of freedom Hamiltonian. The $3D$ solution flux was intercepted by a plan leading to a Poincaré mapping, and specific expressions of the functions led to different mappings.

Chapter 3
A Phenomenological Description
for Chaotic Diffusion

Abstract We discuss a phenomenological description of the chaotic diffusion in this chapter. We use a set of scaling hypotheses leading to the knowledge of three critical exponents demonstrating that the chaotic diffusion is scaling invariant concerning the control parameter, giving the first evidence of a continuous phase transition.

3.1 Dynamical Properties for the Chaotic Sea: A Phenomenological Description

As we discussed in Chap. 2, the mapping of interest is written as

$$\begin{cases} I_{n+1} = I_n + \epsilon \sin(\theta_n), \\ \theta_{n+1} = [\theta_n + \frac{1}{|I_{n+1}|^\gamma}] \mod(2\pi), \end{cases} \tag{3.1}$$

where $\gamma > 0$ and ϵ are control parameters.

As shown in Fig. 2.4, the phase space for the mapping (3.1) has a mixed form incorporating both chaos, a set of periodic islands, and invariant spanning curves. The chaotic diffusion along the phase space has a fascinating property. For small values of I, the term $\frac{1}{|I_{n+1}|^\gamma}$ presumes large values implying that the dynamical variable θ_{n+1} is uncorrelated to θ_n. This uncorrelation allows the dynamical variable θ to behave like a pseudo-random number. Since θ emerges in the argument of the periodic function $\sin(\theta)$, it guides the dynamical variable I to diffuse, consequently growing. As soon as it rises, the ratio $\frac{1}{|I_{n+1}|^\gamma}$ becomes small, bringing a correlation between θ_{n+1} and θ_n and regularity appears in the phase space, including periodic islands and invariant spanning curves.

Unless due to a phase shift, the positive part of the phase space is symmetric concerning the negative part for the dynamical variable I. Such a symmetry yields $\bar{I} = 0$. Hence, the observable $\bar{I}$ is not appropriate for investigating chaotic diffusion. Instead of considering $\bar{I}$, we rather use $\overline{I^2}$ and define $I_{rms} = \sqrt{\overline{I^2}}$. The variable $\overline{I^2}$ is obtained from different kinds of averages

$$\overline{I^2} = \frac{1}{M} \sum_{i=1}^{M} \left[\frac{1}{n} \sum_{j=1}^{n} I_{i,j}^2 \right], \tag{3.2}$$

where n gives the number of iterations for the mapping, and M denotes the ensemble size of different initial conditions. The summation in j provides the average along the orbit, and i gives the average over the ensemble. A set of $M = 1000$ different values for the range of $\theta_0 \in [0, 2\pi]$ was uniformly selected assuming a fixed initial $I_0 = 10^{-3}\epsilon$. With such a choice, the initial conditions are along the chaotic sea and permitted to experience, so far, the highest diffusion as possible. Figure 3.1 shows the behavior of I_{rms} as a function of: (a) n and; (b) $n\epsilon^2$ for three different control parameters.

The curves plotted in Fig. 3.1a are similar while compared among themselves. When the dynamics starts from I_0, each curve grows in a power law for short n and curls towards a regime of saturation for large enough n. The conversion marking the change from the growth regime to the saturation is given by n_x. Each curve grows and saturates at different plateaus obtained for different control parameters. The saturation rises with the increase of ϵ. The curves grow parallel to each other, and an ad hoc transformation $n \to n\epsilon^2$ rescales the regime of growth so that the curves grow together and separate for the saturation at different plateaus as shown in Fig. 3.1b.

The behavior shown in Fig. 3.1 is expected of scaling invariance, commonly observed in phase transitions, and can be described by the subsequent scaling hypotheses:

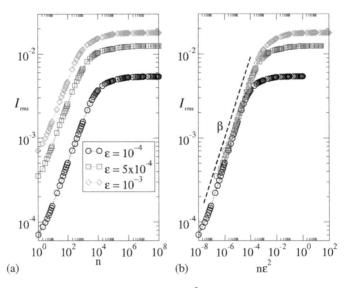

Fig. 3.1 Plot of I_{rms} as a function of: **a** n, and **b** $n\epsilon^2$. The control parameters used were $\gamma = 1$ considering $\epsilon = 10^{-4}$, $\epsilon = 5 \times 10^{-4}$, and $\epsilon = 10^{-3}$, as shown in the figure

1. For $n \ll n_x$, the curves grow as

$$I_{rms} \propto (n\epsilon^2)^\beta, \tag{3.3}$$

where β defines the acceleration exponent and provides a scaling exponent;
2. For $n \gg n_x$, the curves approach a plateau given by

$$I_{rms, \, sat} \propto \epsilon^\alpha, \tag{3.4}$$

where α is the growth exponent;
3. The number of iterations marking the changeover from growth to saturation is written as

$$n_x \propto \epsilon^z, \tag{3.5}$$

where z is also a scaling exponent.

The behavior of I_{rms} can be described using a homogeneous and generalized function written as

$$I_{rms}(n\epsilon^2, \epsilon) = l I_{rms}(l^a n\epsilon^2, l^b \epsilon), \tag{3.6}$$

where l is a scaling factor, a and b are defined as characteristic exponents. The choice of $l^a n\epsilon^2 = 1$ guides to

$$l = (n\epsilon^2)^{-\frac{1}{a}}. \tag{3.7}$$

Replacing this result into Eq. (3.6) gives

$$I_{rms}(n\epsilon^2, \epsilon) = (n\epsilon^2)^{-\frac{1}{a}} I_A((n\epsilon^2)^{-\frac{b}{a}}\epsilon), \tag{3.8}$$

where

$$I_A((n\epsilon^2)^{-\frac{b}{a}}\epsilon) = I_{rms}(1, (n\epsilon^2)^{-\frac{b}{a}}\epsilon) \tag{3.9}$$

is considered constant for $n \ll n_x$. A comparison of Eqs. (3.8) and (3.3) yields $\beta = -1/a$. Numerical simulations give $\beta \cong 1/2$, therefore $a = -2$.
The following choice of $l^b \epsilon = 1$ gives

$$l = \epsilon^{-\frac{1}{b}}. \tag{3.10}$$

Substituting it into Eq. (3.6) leads to

$$I_{rms}(n\epsilon^2, \epsilon) = \epsilon^{-\frac{1}{b}} I_B(\epsilon^{-\frac{a}{b}} n\epsilon^2), \tag{3.11}$$

where the function

$$I_B(\epsilon^{-\frac{a}{b}} n\epsilon^2) = I_{rms}(\epsilon^{-\frac{a}{b}} n\epsilon^2, 1) \tag{3.12}$$

is constant for $n \gg n_x$. Comparing this result with Eq. (3.4) gives $\alpha = -1/b$.

The exponent z can be obtained by a comparison of the two expressions for the scaling factor l given by Eqs. (3.7) and (3.10), leading to $(n\epsilon^2)^\beta = \epsilon^\alpha$. Isolating n, we have

$$n_x = \epsilon^{\frac{\alpha}{\beta}-2}. \tag{3.13}$$

Comparing Eq. (3.13) with Eq. (3.5) gives

$$z = \frac{\alpha}{\beta} - 2. \tag{3.14}$$

The Eq. (3.14) defines a scaling law.

Let us now discuss how to obtain the critical exponents. Indeed, the acceleration exponent is obtained by fitting a power law on the regime of growth of I_{rms}, as represented in Fig. 3.1b. Running numerical simulations for different values of control parameters ϵ and γ, we obtain the exponent $\beta \cong 1/2$, therefore yielding a characteristic exponent $a = -2$.

A procedure to obtain the critical exponent α is made for $n \gg n_x$ and gives the asymptotic behavior for I_{rms}. It was assumed γ as fixed and varied the parameter ϵ. The range of interest letting to define the transition from integrability to non-integrability is given for $\epsilon \in [10^{-4}, 10^{-2}]$. Figure 3.2a illustrates a plot of $I_{rms,\,sat}$ versus ϵ for the control parameter $\gamma = 1$. The exponent α was obtained numerically after a power law fitting as $\alpha = 0.508(4)$. Considering $\gamma = 2$ we obtained $\alpha = 0.343(2)$, as shown in Fig. 3.2b.

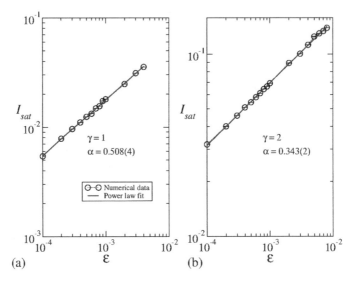

Fig. 3.2 Plot of $I_{rms,\,sat}$ versus ϵ for: **a** $\gamma = 1$ and **b** $\gamma = 2$. The critical exponents obtained are: **a** $\alpha = 0.508(4)$ and **b** $\alpha = 0.343(2)$

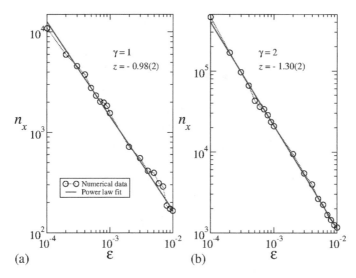

Fig. 3.3 Plot of n_x versus ϵ for: **a** $\gamma = 1$ and **b** $\gamma = 2$. The critical exponents obtained are: **a** $z = -0.98(2)$ and **b** $z = -1.30(2)$

The procedure to obtain the critical exponent z is made from the plot of n_x versus ϵ considering a constant value of γ. The crossover n_x specifies a changeover point where the I_{rms} curve changes from the growth regime to a plateau. Fig. 3.3a gives the behavior of n_x versus ϵ for $\gamma = 1$. A power law fitting furnishes $z = -0.98(2)$. For $\gamma = 2$, we obtained $z = -1.30(2)$, as displayed in Fig. 3.3b. Both results agree with Eq. (3.14).

Let us apply the scaling transformations $I_{rms} \rightarrow I_{rms}/\epsilon^{\alpha}$ and $n \rightarrow n/\epsilon^z$. They are obtained from distinct control parameters allowing to overlay all curves of I_{rms} onto a universal plot, demonstrating the diffusion along the chaotic sea is scaling-invariant concerning the control parameter. This is a sign of a transition from integrability to non-integrability. This behavior can be seen in Fig. 3.4, confirming the scaling invariance of the chaotic sea concerning the variation of the control parameter ϵ.

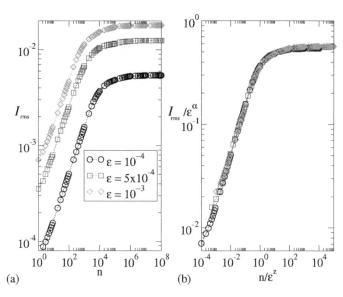

Fig. 3.4 a Plot of I_{rms} versus n for $\gamma = 1$ and different values of ϵ as shown in the figure. **b** Overlap of the curves shown in **a** onto a single and hence universal plot after the scaling transformations $I_{rms} \rightarrow I_{rms}/\epsilon^{\alpha}$ and $n \rightarrow n/\epsilon^{z}$

3.2 Summary

In this chapter, we studied chaotic diffusion considering a phenomenological description and used a set of scaling hypotheses. They led to a scaling law where three critical exponents were obtained numerically. The scaling invariance observed through overlapping all curves of I_{rms} onto a single and hence universal plot is a common feature present in continuous phase transitions.

Chapter 4
A Semi-phenomenological Description for Chaotic Diffusion

Abstract We discuss a semi-phenomenological procedure for the chaotic diffusion in this chapter leading to an estimation for the critical exponents α, β, and z. The exponent, α, is obtained using a connection with the standard mapping in a transition from locally to globally chaotic dynamics, where invariant spanning curves are destroyed. The exponent β is obtained by transforming the equation of differences into a differential equation, allowing a prompt solution. The critical exponent z is obtained by using the scaling law.

4.1 A Semi-phenomenological Approach

In Chap. 3, we described the behavior of the chaotic diffusion utilizing a set of scaling hypotheses guiding to a scaling law. The mapping considered is written as

$$\begin{cases} I_{n+1} = I_n + \epsilon \sin(\theta_n), \\ \theta_{n+1} = [\theta_n + \frac{1}{|I_{n+1}|^\gamma}] \ \mathrm{mod}(2\pi), \end{cases} \tag{4.1}$$

where $\gamma > 0$ and ϵ are control parameters.

We now discuss a procedure that transforms the equation of differences of the mapping into a differential equation that can be integrated, leading to an analytical approximation of the critical exponent β. The exponent α is obtained via the localization of the first invariant spanning curve that plays a significant role in the scaling of the chaotic diffusion. The exponent z can be obtained by using the scaling law, as discussed in the previous chapter, or by considering a direct integration of the transformed equation using explicit limits for the integration.

As shown in the phase space, the chaotic sea is limited by the first invariant spanning curve on either side of the action axis I. The phase space is characterized chiefly, but not only by regular regions above the first invariant spanning curve. Below the curve, large portions of chaos are observed surrounding periodic islands. The first invariant spanning curve separates two different regions: (i) above the curve where local chaos can be observed and; (ii) below the curve where global chaos dominates the dynamics.

The transition from locally chaotic to globally chaotic dynamics is observed in the standard mapping for a critical control parameter. The mapping is written as

$$\begin{cases} I_{n+1} = I_n + k \sin(\theta_n), \\ \theta_{n+1} = [\theta_n + I_{n+1}] \mod(2\pi), \end{cases} \qquad (4.2)$$

where k is a control parameter. For $k = 0$, the mapping is integrable, and for $k \neq 0$, the mapping is non-integrable. The phase space is mixed for small values of k and shows a transition from local chaos for $k < 0.9716\ldots$ to global chaos with $k \geq 0.9716\ldots$. At this point and beyond, the last invariant spanning curve is destroyed.

To use this property in the mapping (2.16), we assume near the invariant spanning curve, the variable I is described as

$$I_n = I + \Delta I_n, \qquad (4.3)$$

where I corresponds to a characteristic value along the invariant spanning curve and ΔI_n represents a small perturbation of I. The first equation of (2.16) is written as

$$\Delta I_{n+1} = \Delta I_n + \epsilon \sin(\theta_n). \qquad (4.4)$$

Moreover, the second equation of (2.16) is given by

$$\begin{aligned} \theta_{n+1} &= \theta_n + \frac{1}{(I + \Delta I_{n+1})^\gamma} \\ &= \theta_n + \frac{1}{I^\gamma} \left(1 + \frac{\Delta I_{n+1}}{I} \right)^{-\gamma}. \end{aligned} \qquad (4.5)$$

Expanding Eq. (4.5) in Taylor series and assuming first-order terms in $\Delta I_{n+1}/I$ we have

$$\begin{aligned} \theta_{n+1} &= \theta_n + \frac{1}{I^\gamma} \left(1 - \gamma \frac{\Delta I_{n+1}}{I} \right) \\ &= \left[\theta_n + \frac{1}{I^\gamma} - \frac{\gamma \Delta I_{n+1}}{I^{\gamma+1}} \right]. \end{aligned} \qquad (4.6)$$

To make a connection with the standard mapping, Eq. (4.4) must be multiplied by $-\gamma/I^{\gamma+1}$ and added by the term $1/I^\gamma$. The subsequent variables are defined

$$J_n = \frac{1}{I^\gamma} - \frac{\gamma \Delta I_n}{I^{\gamma+1}}, \qquad (4.7)$$

$$\phi_n = \theta_n + \pi. \qquad (4.8)$$

The mapping (2.16) is written near the invariant spanning curve as

$$\begin{cases} J_{n+1} = J_n + \frac{\gamma \epsilon}{I^{\gamma+1}} \sin(\phi_n), \\ \phi_{n+1} = [\phi_n + J_{n+1}] \mod(2\pi), \end{cases} \qquad (4.9)$$

The mapping (4.9) has control parameters that can be grouped as

$$K_{ef} = \frac{\gamma\epsilon}{I^{\gamma+1}}. \tag{4.10}$$

Near the invariant spanning curve, we have $K_{ef} \cong 0.9716\ldots$. Therefore, the localization for the first invariant spanning curve in the phase space is given by

$$I_{Fisc} = \left[\frac{\gamma\epsilon}{K_{ef}}\right]^{\frac{1}{\gamma+1}}$$

$$= \left[\frac{\gamma}{K_{ef}}\right]^{\frac{1}{\gamma+1}} \epsilon^{\frac{1}{\gamma+1}}. \tag{4.11}$$

The first invariant spanning curve's localization defines the chaotic region's upper limit. Then, I_{Fisc} limits the size of I_{rms}. Indeed the numerical value of $I_{rms,\,sat}$, that is observed for $n \gg n_x$, is defined by a fraction of I_{Fisc}. Therefore I_{Fisc} defines a law that I_{rms} obeys as a function of ϵ. An immediate comparison of Eq. (4.11) with Eq. (3.4) yields

$$\alpha = \frac{1}{\gamma + 1}. \tag{4.12}$$

Equation (4.12) defines a relation between the critical exponent α and the control parameter γ.

Let us now describe the exponent β. To do so, we use the first equation of (2.16). Squaring both sides, we have

$$I_{n+1}^2 = I_n^2 + 2\epsilon I_n \sin(\theta_n) + \epsilon^2 \sin^2(\theta_n). \tag{4.13}$$

Doing an ensemble average in Eq. (4.13) with $\theta \in [0, 2\pi]$ and considering that in the chaotic domains, the two dynamical variables I and θ are statistically independent yields

$$\overline{I^2}_{n+1} = \overline{I^2}_n + \frac{\epsilon^2}{2}, \tag{4.14}$$

since that $\overline{\sin(\theta)} = 0$ and $\overline{\sin^2(\theta)} = 1/2$. Considering the case where ϵ is sufficiently small, $\overline{I^2}_{n+1} - \overline{I^2}_n$ is small, we can use the following approximation:

$$\overline{I^2}_{n+1} - \overline{I^2}_n = \frac{\overline{I^2}_{n+1} - \overline{I^2}_n}{(n+1) - n}$$

$$\cong \frac{d\overline{I^2}}{dn} = \frac{\epsilon^2}{2}. \tag{4.15}$$

A first-order differential equation can be integrated easily, hence

$$\int_{I_0}^{I(n)} d\overline{I^2} = \int_0^n \frac{\epsilon^2}{2} dn', \tag{4.16}$$

leading to

$$\overline{I^2}(n) = I_0^2 + \frac{\epsilon^2}{2}n. \tag{4.17}$$

Applying square root on both sides gives

$$I_{rms} = \sqrt{I_0^2 + \frac{\epsilon^2}{2}n} . \tag{4.18}$$

In the limit of small values of I_0, the expression is written as

$$I_{rms} \cong \frac{1}{\sqrt{2}}(n\epsilon^2)^{\frac{1}{2}}. \tag{4.19}$$

We can then compare Eq. (3.3), allowing us to conclude that $\beta = 1/2$ is in good agreement with the numerical simulations. The term $n\epsilon^2$ appears naturally in Eq. (4.19).

The critical exponent z can also be obtained analytically. To do that, we have to consider the integration limits of Eq. (4.16) as $I_0 \to 0$ and $I(n) = I_{Fisc}$. In such limits, we obtain an approximation for n_x as

$$\left[\frac{\gamma\epsilon}{K_{ef}}\right]^{\frac{1}{\gamma+1}} = \left[\frac{n_x\epsilon^2}{2}\right]^{\frac{1}{2}}, \tag{4.20}$$

and that isolating n_x gives

$$n_x = 2\left[\frac{\gamma}{K_{ef}}\right]^{\frac{2}{\gamma+1}} \epsilon^{\frac{-2\gamma}{\gamma+1}}. \tag{4.21}$$

Equation (4.21) can be compared with Eq. (3.5), hence leading to the equation

$$z = -\frac{2\gamma}{\gamma+1}. \tag{4.22}$$

The analytical expressions for the exponents α, β, and z obtained in this section agree with the numerical simulations.

4.2   Summary

In this chapter, we applied a transformation to the equation of differences into a differential equation in which the solution gives one of the critical exponents, the acceleration exponent. The second critical exponent is obtained from the localization of the first invariant spanning curve by using a connection with a specific transition observed in the standard mapping.

Chapter 5
A Solution of the Diffusion Equation

Abstract This chapter is dedicated to the solution of the diffusion equation. It gives the probability density to observe a particle with a given action at a specific time in the phase space. It is fundamental to investigate the chaotic diffusion along the phase space. We impose particular boundary conditions and concentrate all the particles leaving from an initial action and resolve analytically the probability density that provides the probability a particle can be observed with action $I \in [-I_{fisc}, I_{fisc}]$ at any time n. The knowledge of the probability density furnishes all the relevant observables, including the scaling invariance of the chaotic diffusion.

5.1   A Solution of the Diffusion Equation

In this section, we consider a rather different approach than the ones discussed in the two previous chapters. We obtain the probability density of observing a particle in the chaotic region with an action $I \in [-I_{fisc}, I_{fisc}]$ at a given instant of time n, where I_{fisc},[1] resembles the first invariant spanning curve. We solve the diffusion equation imposing specific boundary and initial conditions to obtain the probability $P(I, n)$. The observables obtained along the chaotic sea correspond to the momenta of the distribution and give either the average value $\bar{I} = \int_{-I_{fisc}}^{I_{fisc}} I P(I, n) dI$ and likewise the average quadratic value given by $\bar{I^2} = \int_{-I_{fisc}}^{I_{fisc}} I^2 P(I, n) dI$, from where $I_{rms}(n) = \sqrt{\bar{I^2}(n)}$ is obtained.

A careful look at the phase space shown in Fig. 2.4 allows one to see that the dynamics experienced for particles along the chaotic region resembles a normal diffusion process. It implies that from a pair of initial conditions (I_0, θ_0), the time evolution of $I(n)$ can be given in a particular way such that there exists a probability of the dynamics to raise the value of I with a probability p. At the same time, it may decrease with probability q such that $p + q = 1$. Accordingly, the probability of observing a particle in the chaotic region restricted to move between the two invariant spanning curves $I \in [-I_{fisc}, I_{fisc}]$ is $P(I, n)$ and that is obtained from the

---

[1] The sub-index $fisc$ symbolizes first invariant spanning curve.

solution of the diffusion equation, given by

$$\frac{\partial P(I,n)}{\partial n} = D\frac{\partial^2 P(I,n)}{\partial I^2},\tag{5.1}$$

where D is the diffusion coefficient. A boundary condition for $P(I, n)$ at the invariant spanning curves is $\frac{\partial P(I,n)}{\partial I}\Big|_{\pm I_{fisc}} = 0$. It insinuates there is no particle flux through the invariant spanning curves, guaranteeing the particle is always bounded to move along the chaotic dynamics.

To solve Eq. (5.1), it is assumed the function $P(I, n)$ is composed as $P(I, n) = X(I)N(n)$, where the function $X(I)$ depends only on I, while $N(n)$ is a function that depends only on n. Applying the boundary and initial conditions, the expression for the probability density assumes the form

$$P(I,n) = \frac{1}{2I_{fisc}} + \frac{1}{I_{fisc}}\sum_{k=1}^{\infty}\cos\left(\frac{k\pi I}{I_{fisc}}\right)e^{-\frac{k^2\pi^2 Dn}{I_{fisc}^2}}.\tag{5.2}$$

The first term on the right-hand side gives the stationary state distribution, while the second term is related to the transient dynamics.

The diffusion coefficient D is obtained from $D = \overline{\Delta I^2}/2 = \frac{\overline{I^2}_{n+1}-\overline{I^2}_n}{2}$. Using the first equation of mapping (2.16), we obtain $D = \frac{\epsilon^2}{4}$. D is a constant by supposing statistical independence between the dynamical variables I and θ in the chaotic sea. It works well when the dynamics passes relatively far away from the periodic regions responsible for producing local trappings. Such a dynamical regime affects the transport of particles, therefore, modifying the expression of D.

The variable $\overline{I^2}(n)$ is obtained from $\overline{I^2}(n) = \int_{-I_{fisc}}^{I_{fisc}} I^2 P(I,n)dI$, leading to

$$\overline{I^2}(n) = I_{fisc}^2\left[\frac{1}{3} + \frac{4}{\pi^2}\sum_{k=1}^{\infty}\frac{(-1)^k}{k^2}e^{-\frac{k^2\pi^2 Dn}{I_{fisc}^2}}\right].\tag{5.3}$$

Compare the curve obtained from Eq. (5.3) with the one generated from Eq. (3.2). The latter is generated by considering two different averages in its calculation. One is the average over an ensemble of different initial conditions, while the other is an average over time. Nonetheless, Eq. (5.3) considers only the ensemble average. The implementation of the time average considers $<\overline{I^2}(n)> = \frac{1}{n}\sum_{j=1}^{n}\overline{I^2}_j(n)$. Since only the last term is independent of n, the expression assumes the form

$$\frac{1}{n}\sum_{j=1}^{n}e^{-\frac{jk^2\pi^2 D}{I_{fisc}^2}} = \frac{1}{n}\left[e^{-\frac{k^2\pi^2 D}{I_{fisc}^2}} + e^{-\frac{2k^2\pi^2 D}{I_{fisc}^2}} + \cdots + e^{-\frac{nk^2\pi^2 D}{I_{fisc}^2}}\right]$$

$$= \frac{1}{n}\left[\frac{1 - e^{-\frac{nk^2\pi^2 D}{I_{fisc}^2}}}{1 - e^{-\frac{k^2\pi^2 D}{I_{fisc}^2}}}\right].\tag{5.4}$$

Then the expression of $I_{rms}(n) = \sqrt{<\overline{I^2}(n)>}$ is written as

$$I_{rms}(n) = I_{fisc} \sqrt{\frac{1}{3} + \frac{4}{\pi^2} \sum_{k=1}^{\infty} \frac{(-1)^k}{k^2} e^{-\frac{k^2\pi^2 D}{I_{fisc}^2}} \frac{1}{n} \left[\frac{1 - e^{-\frac{nk^2\pi^2 D}{I_{fisc}^2}}}{1 - e^{-\frac{k^2\pi^2 D}{I_{fisc}^2}}}\right]}. \tag{5.5}$$

The saturation limit is obtained when $n \to \infty$ leads to

$$I_{sat} = \frac{I_{fisc}}{\sqrt{3}}.$$

Figure 5.1 shows a plot of $I_{rms}(n)$ versus n for different control parameters, as shown in the own figure. The symbols correspond to the numerical simulation, while the continuous curves correspond to the solution of Eq. (5.5) with the same control parameters used in the numerical simulations.

Let us now discuss the dominant terms of Eq. (5.5) by considering a first-order Taylor expansion using $k = 1$ for $I_0 \cong 0$, that gives

$$\cos\left(\frac{\pi I_0}{I_{fisc}}\right) \cong \left[1 - \frac{1}{2}\left(\frac{\pi I_0}{I_{fisc}}\right)^2\right], \tag{5.6}$$

$$e^{-\frac{\pi^2 D}{I_{fisc}^2}} \cong 1 - \frac{\pi^2 D}{I_{fisc}^2}, \tag{5.7}$$

$$1 - e^{-\frac{\pi^2 D}{I_{fisc}^2}} \cong \frac{\pi^2 D}{I_{fisc}^2}. \tag{5.8}$$

Fig. 5.1 Plot of $I_{rms}(n)$ versus n for different control parameters. The symbols denote the numerical simulations, and the continuous curves correspond to Eq. (5.5) with the same control parameters as used in the numerical simulations

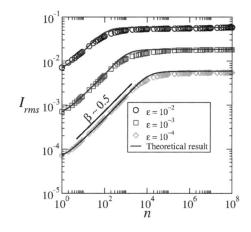

Due to the existence of the term $\frac{1}{n}$ in Eq. (5.5), the expression for the exponential in n must be expanded until second order, hence

$$1 - e^{-\frac{\pi^2 Dn}{I_{fisc}^2}} \cong \frac{\pi^2 D}{I_{fisc}^2} - \frac{\pi^4 D^2 n^2}{2 I_{fisc}^4}. \tag{5.9}$$

Substituting these terms in Eq. (5.5), considering that $D = \frac{\epsilon^2}{4}$ yields

$$I_{rms}(n) \cong \sqrt{\left(\frac{1}{3} - \frac{4}{\pi^2}\right) \epsilon^{\frac{1}{1+\gamma}} + \frac{\epsilon^2 n}{2}}, \tag{5.10}$$

that due to the variation in n leads to $I_{rms} \propto \sqrt{\frac{\epsilon^2 n}{2}}$. This result confirms the exponent $\beta = 1/2$ giving support also to the ad hoc transformation $n \to n\epsilon^2$. When the regime of growth intercepts the saturation, the crossover iteration number is given by

$$n_x \propto \frac{2}{3} \epsilon^{-\frac{2\gamma}{1+\gamma}}, \tag{5.11}$$

hence leading to the critical exponent $z = -\frac{2\gamma}{1+\gamma}$.

5.2   Summary

In this chapter, we considered an analytical solution of the diffusion equation, which supplies the probability of observing a given particle with action I at a specific time n. From the probability, the expressions of the momenta of the distribution are obtained, permitting a description of I_{rms} versus n analytically. The application of explicit limits yields in the determination of the three critical exponents α, β, and z.

Chapter 6
Characterization of a Continuous Phase Transition in an Area-Preserving Map

Abstract We discuss along the chapter four essential points to characterize a transition from integrability to non-integrability in a two-dimensional, nonlinear, and area-preserving mapping. A parameter ϵ controls the transition and is closely related to the order parameter. The average squared action along the chaotic sea presents a scaling invariance concerning the control parameter. This property is a characteristic of a continuous phase transition. Therefore, the transition from integrability to non-integrability is similar to the second order, also called continuous phase transition. A property of such transition is that the order parameter approaches zero at the same time the response of the order parameter to the conjugate field (susceptibility) diverges.

6.1 Evidence of a Phase Transition

As we discussed in previous chapters, for a transition, we want to discuss a Hamiltonian that can be written as $H(I_1, \theta_1, I_2, \theta_2) = H_0(I_1, I_2) + \epsilon H_1(I_1, \theta_1, I_2, \theta_2)$ where H_0 corresponds to the integrable part and H_1 denotes the non integrable part with ϵ controlling a transition. The system is integrable for the case of $\epsilon = 0$ since both energy and action are preserved. The case of $\epsilon \neq 0$ breaks the integrability since only the energy is preserved. The transition contains elements of a continuous phase transition, given the order parameter goes continuously to zero at the same time that the response of the order parameter to the external perturbation also called susceptibility, diverges at the same limit.

The mapping is given by

$$\begin{cases} I_{n+1} = I_n + \epsilon \sin(\theta_n), \\ \theta_{n+1} = [\theta_n + \frac{1}{|I_{n+1}|^{\gamma}}] \mod(2\pi). \end{cases} \qquad (6.1)$$

Indeed, when the action I is small, the variable θ_n is uncorrelated with θ_{n+1} conducting to the diffusion of chaotic orbits, therefore, allowing I to increase. As soon as I grows, the angular variables present correlation, and regularity emerges in the phase space, leading to periodic islands and invariant spanning curves, which play a significant role in the dynamics.

We address a scaling present in the transition from integrability to non-integrability, focusing on the discussion of four principal items to analyze a phase transition: (1) Identify the broken symmetry; (2) Define the order parameter; (3) Discuss the elementary excitation; and (4) Discuss the topological defects which impact in the transport of particles.

6.2 Break of Symmetry

We notice that ϵ controls an interesting part of the dynamics. For $\epsilon = 0$, the system is integrable since both the energy E and action I are constants. The integrable dynamics leads to a foliated phase space with constant I. The system's dynamics for $\epsilon = 0$ is trivial and entirely foreseen, and no exponentially spreading of two nearby initial conditions is observed. When $\epsilon \neq 0$, the phase space is no longer foliated, assuming a mixed form. In this new configuration, the phase space has more complicated dynamics. Depending on the initial conditions and control parameters, a large chaotic sea is limited by invariant spanning curves and surrounds periodic islands. Because of the area preservation, the stable islands do not let particles moving along the chaotic sea move in. Simultaneously, do not permit particles inside to escape through the islands. Consequently, they compare topological defects violating the ergodicity condition for the phase space. The invariant spanning curves also have a crucial function for the dynamics. Since they obstruct the passage of particles from downside to upside and vice versa, they define the size of the chaotic sea. Figure 6.1

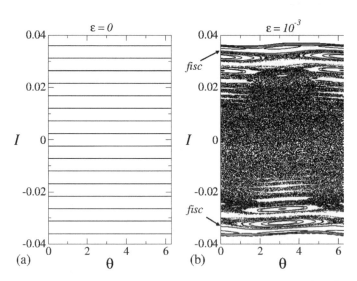

Fig. 6.1 Plot of the phase space for the mapping (6.1) for: **a** $\epsilon = 0$ and **b** $\epsilon = 10^{-3}$. The curves shown in **b** correspond to the first invariant spanning curves and scale with $\epsilon^{1/(1+\gamma)}$

shows a plot of the phase space for the mapping (6.1) considering two different control parameters, namely: (a) $\epsilon = 0$ and (b) $\epsilon = 10^{-3}$.

The discussion presented in the earlier paragraph allows us to think about what is the broken symmetry the system has passed. Firstly, for $\epsilon = 0$, the phase space is foliated, therefore completely regular. Each curve shown in Fig. 6.1a depends only on the initial action, which is preserved along the dynamics. Since the dynamics preserves the action, two different and nearby initial conditions do not disperse from each other exponentially as time goes on, a condition requested for chaotic dynamics. Therefore, we observe regular dynamics on the phase space characterizing then a phase of regularity for the dynamics. On the other hand, when $\epsilon \neq 0$ the nonlinear function $\sin(\theta)$ affects the time evolution of the particles, affecting the dynamics and destroying the regularity present in the phase space. It exhibits a mixed format and possesses periodic dynamics, characterized by the different period fixed points in the middle of the periodic islands, invariant spanning curves marked by continuous curves in Fig. 6.1b, and a chaotic sea. Along the chaotic sea, two nearby initial conditions are apart from each other exponentially, as requested to characterize chaotic dynamics. The chaotic sea current in the phase space births with a length of size well defined in action. An initial condition offered in the chaotic sea spreads over the phase space along the range $I \in \left(-\left[\frac{\gamma\epsilon}{0.9716...}\right]^{1/(1+\gamma)}, \left[\frac{\gamma\epsilon}{0.9716...}\right]^{1/(1+\gamma)}\right)$. The invariant spanning curves define the positive and negative limits for the chaotic sea, which work as barriers that do not let the particles cross through. The collapse of the regularity marks the broken symmetry for $\epsilon \neq 0$ and defines the window size for the chaotic dynamics. Therefore, the broken symmetry is from regularity to chaotic dynamics and hence to chaotic diffusion.

We also notice that the mixed phase space with islands and invariant spanning curves exhibits different averages for chaotic diffusion while measured for an ensemble of different initial conditions and measured in time. Therefore, time average is different from the microcanonical average. Because of such a difference, the basic assumption of ergodicity is broken. Hence ergodicity is not observed in such a system. Moreover, parts of the phase space can not intermingle with others, such as chaotic dynamics can not invade the periodic structures and vice versa. As a particle moving in chaotic dynamics passes close enough to the regular structures or on the invariant spanning curves, it may stay locally trapped in a dynamics which is called stickiness.

Moreover, from the equations of the mapping (6.1), we notice the first equation depends on ϵ, namely $I_{n+1} = I_n + \epsilon \sin(\theta_n)$. Of course, when $\epsilon = 0$, the equation is symmetric, hence $I_{n+1} = I_n$. It then gives the second quantity, among the energy, which is constant. The case of $\epsilon \neq 0$ breaks the symmetry of the equation and of the phase space, leading to mixed dynamics, therefore an algebraic break of symmetry.

6.3 Order Parameter

Allow us now to argue on a conceivable order parameter. The dynamics is regular for $\epsilon = 0$. Chaos emerges for $\epsilon \neq 0$ and specific initial conditions. Because of the existence of the two invariant spanning curves, one from positive and the other from opposing sides, the chaotic diffusion is limited. Due to the symmetry of the phase space, the average action is not an excellent variable to investigate the chaotic diffusion. Instead, a good one is a root mean square of the squared action, and its value for a long enough time gives the saturation of the chaotic diffusion and is written as $I_{sat} \propto \epsilon^{\alpha}$. This variable is a good candidate as an order parameter since it goes continuously to zero as $\epsilon \to 0$, marking an ordered phase. Nevertheless, it differs from zero when $\epsilon \neq 0$ and leads to chaotic diffusion on the phase space.

Let us compare this transition with the one observed in a ferromagnetic system. Consider a system composed of spins that can interact with any other and align with an external field. The spontaneous magnetization m is the order parameter in such a system. Nevertheless, local interactions define the system's magnetization for the null external field, dependent on the external temperature T. For a temperature T below a critical one T_c, non-null magnetization is observed. With the temperature increase above T_c, the order phase marked by aligned spins with each other is destroyed, and null magnetization is observed. As soon as $T \to T_c$ from below, the magnetization goes smoothly and continually to zero. The response of the order parameter to the external field gives the magnetic susceptibility χ, which diverges in such a limit.

These two premises are elements of a continuous phase transition. Back to the chaotic model, as soon as the control parameter ϵ is made different from zero, the chaotic sea births with restricted size. Figure 6.2 shows the positive Lyapunov exponent for a large window of variation of the control parameter $\epsilon \in [10^{-6}, 10^{-2}]$. We notice that the positive Lyapunov exponent λ varies very little, typically along the range $\lambda \in [1.5, 1.75]$ as compared to the large range of variation of the control parameter $\epsilon \in [10^{-6}, 10^{-2}]$. It allows us to suppose that the chaotic sea births indeed with a limited size and the chaotic dynamics has a finite positive Lyapunov exponent. This almost constant value is mostly associated with the chaotic sea scaling invariant concerning the control parameter ϵ. Therefore, even though the Lyapunov exponent characterizes chaotic dynamics, it is not a good observable to define an order parameter.

Let us now discuss the natural observable along the chaotic sea to demonstrate diffusion, which is the square root of the averaged squared action. The behavior of I_{rms} can be made as follows, as shown in Fig. 6.3a. For an initial action typically $I_0 \cong 0$, the curves of $I_{rms} \propto (n\epsilon^2)^{\beta}$ grow with the exponent $\beta \cong 1/2$ yielding the diffusion of particles to be equivalent to normal diffusion.

For large enough time and due to the presence of invariant spanning curves, the curve of $I_{rms,sat} \propto \epsilon^{\alpha}$ with $\alpha = \frac{1}{1+\gamma}$. The regime marking the change of growth to

Fig. 6.2 Plot of the positive Lyapunov exponent for a large range of control parameter $\epsilon \in [10^{-6}, 10^{-2}]$

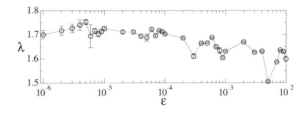

Fig. 6.3 a Plot of different curves of I_{rms} versus n for the control parameters and initial action as labeled in the figure. **b** Overlap of the curves shown in **a** onto a single and universal plot

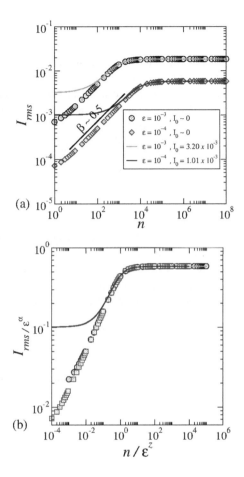

the saturation is given by $n_x \propto \epsilon^z$ with $z = -\frac{2\gamma}{\gamma+1}$. The curves overlap well after appropriate scaling transformation, as shown in Fig. 6.3b. The overlap of the curves shown in Fig. 6.3b confirms a scaling invariance observed for chaotic dynamics near a transition from integrability to non-integrability for the mapping (6.1). We also notice a scaling in the curves even if the initial action is not small enough, as in the continuous curves.

In the limit of $\epsilon \to 0$, the order parameter $I_{sat} \propto \epsilon^\alpha$ approaches zero continuously. The theory of second-order phase transition says that the susceptibility, that is, the response of the order parameter to the related parameter ϵ, must diverge in the above limit. The susceptibility is calculated as

$$\chi = \frac{\partial I_{sat}}{\partial \epsilon} = \left[\frac{1}{1+\gamma}\right] \frac{1}{\epsilon^{\frac{\gamma}{1+\gamma}}}. \tag{6.2}$$

Since γ is a non-negative number, in the limit of $\epsilon \to 0$ then $\chi \to \infty$. That is a clear sign of a second-order phase transition.

6.4 Topological Defects

Let us now discuss the topological defects. This terminology is imported from statistical mechanics, meaning the reasons for the break of ergodicity in the dynamics. They may be associated with dislocation in an array of atoms, which is caused by an extra column of atoms in a material. Nonetheless, they can be observed in magnetic material where a magnetization from a specific domain alines out from a point in all directions. Moreover, they exist in nematic crystals, among many other types. In the present section, the defects are connected with regions of the phase space where the diffusion is affected by local trapping.

In a conservative system where the phase space has mixed form, chaos, periodic islands, and invariant tori are all observed. Each one of them is obtained by evolving different initial conditions. However, a dynamical regime called stickiness directly affects the transport of particles. Stickiness is observed when a particle moving along a chaotic domain passes close enough to an island or an invariant curve and stays moving around, hence temporarily confined. The confinement may be short or not, in case the diffusion is affected.

For the family of area-preserving mapping, written as

$$\begin{cases} I_{n+1} = I_n + \epsilon \sin(\theta_n), \\ \theta_{n+1} = [\theta_n + \frac{1}{|I_{n+1}|^\gamma}] \mod(2\pi), \end{cases} \tag{6.3}$$

where $\gamma > 0$ is the control parameter, the fixed points are obtained by imposing the following conditions

$$I_{n+1} = I_n = I, \tag{6.4}$$

$$\theta_{n+1} = \theta_n = \theta + 2m\pi, \quad m = 1, 2, 3 \ldots . \tag{6.5}$$

The solution for Eqs. (6.4) and (6.5), applied to the mapping (6.3) gives the fixed points

$$(\theta, I) = \begin{cases} 0, \pm \left(\frac{1}{2m\pi}\right)^{\frac{1}{\gamma}}, \\ \pi, \pm \left(\frac{1}{2m\pi}\right)^{\frac{1}{\gamma}}. \end{cases} \tag{6.6}$$

The classification for an elliptic fixed point is given for both $(0, (1/(2m\pi))^{1/\gamma})$ and $(\pi, -(1/(2m\pi))^{1/\gamma})$ when

$$m < \frac{1}{2\pi} \left[\frac{4}{\epsilon\gamma}\right]^{\frac{\gamma}{\gamma+1}}. \tag{6.7}$$

For

$$m \geq \frac{1}{2\pi} \left[\frac{4}{\epsilon\gamma}\right]^{\frac{\gamma}{\gamma+1}}, \tag{6.8}$$

the fixed points $(0, -(1/(2m\pi))^{1/\gamma})$ and $(\pi, (1/(2m\pi))^{1/\gamma})$ are hyperbolic, therefore unstable.

Around each of the elliptic fixed points, there is an island of stability. When a particle moves in the dynamics dictated by the mapping (6.3) and passes near enough to the islands, it becomes locally trapped. If the system's dynamics were chaotic with no periodic points and the average over a microcanonical ensemble is equal to the time average, the system would be ergodic. It is not the case since periodic islands exist in the phase space, which is mixed. The islands are interpreted as equivalent to topological defects destroying the ergodicity of the system. When the dynamics pass near the islands, stickiness is observed, modifying the probability of a particle surviving or escaping from a given region.

6.5 Elementary Excitations

This section is devoted to discussing elementary excitation. The parameter ϵ present as a prefactor of the nonlinear function $\sin(\theta_n)$ defines the elementary excitation of the dynamics. For the chaotic dynamics and assuming statistical independence of the dynamical variables θ and I and for small values of I, the first equation of mapping (6.1) leads to an equivalent random walk dynamics with the average size of $\epsilon/\sqrt{2}$. It then turns out to be the elementary excitation of the system. Taking the square of the first equation of mapping (6.1), making an average over an ensemble of various initial phases $\theta_0 \in [0, 2\pi]$ and accepting statistical independence between I and θ, we obtain $\overline{I^2}_{n+1} = \overline{I^2}_n + \frac{\epsilon^2}{2}$. This equation also allows obtaining the diffusion coefficient as $D = \frac{\epsilon^2}{4}$. A transformation of the equation of differences into a differential equation assumes that $\overline{I^2}_{n+1} - \overline{I^2}_n = (\overline{I^2}_{n+1} - \overline{I^2}_n)/((n+1) - n) \cong \frac{d\overline{I^2}}{dn} = \frac{\epsilon^2}{2}$, then it leads to the following result $\overline{I^2}(n) = \overline{I^2}_0 + n\frac{\epsilon^2}{2}$. It gives an excitation that scales with ϵ.

6.6 Summary

We have investigated the basic elements used to identify and classify a second-order phase transition. The scaling present in the chaotic diffusion is linked to the chaotic domain's limit size, leading to a set of critical exponents used to transform the curves of I_{rms} onto a universal curve. The order parameter was specified as $I_{sat} \propto \epsilon^\alpha$ with $\alpha = \frac{1}{1+\gamma}$ where ϵ is the control parameter and that $I_{sat} \to 0$ when $\epsilon \to 0$. The suscep- tibility $\chi = \frac{1}{1+\gamma} \frac{1}{\epsilon^{\frac{2}{1+\gamma}}}$ diverges in the limit of $\epsilon \to 0$. These two results are signatures of continuous phase transitions. The nonlinear function produces elementary exci- tations, leading the dynamics at the low action domain to behave as a random walk particle. Finally, the existence of the periodic islands was interpreted as topologi- cal defects in the phase space that modify the system's transport properties, leading to sticky dynamics. The discussion presented here allows us to conclude the phase transition from integrability to non-integrability in the mapping (6.1) is analogous to a second-order phase transition.

Chapter 7
Scaling Invariance for Chaotic Diffusion in a Dissipative Standard Mapping

Abstract We describe a scaling invariance for chaotic orbits in a dissipative standard mapping by using a phenomenological approach. Two scaling laws are derived, and a set of five critical exponents are obtained. The observable of interest demonstrating a scaling invariance is the average squared action for chaotic orbits in the phase space.

7.1 A Dissipative Standard Mapping

In this section, we investigate the behavior of chaotic diffusion for a dissipative standard mapping. The equations describing the mapping are written as

$$\begin{cases} I_{n+1} = (1 - \gamma)I_n + \epsilon \cos(\theta_n), \\ \theta_{n+1} = (\theta_n + I_{n+1}) \bmod(2\pi), \end{cases} \tag{7.1}$$

where $\gamma \in [0, 1]$ is the dissipative parameter, and ϵ resembles the intensity of the nonlinearity.

This system has two well-known transitions for $\gamma = 0$ (conservative case). The first is a transition from integrability for $\epsilon = 0$ to non-integrability when $\epsilon \neq 0$. The phase space is foliated when $\epsilon = 0$ and shows a mixed structure containing periodic islands, chaotic seas, and invariant spanning curves. The curves limit the diffusion of chaotic particles to a closed region. A second one happens at a critical value of $\epsilon_c = 0.9716\ldots$. At this parameter, the system admits a transition from local chaos when $\epsilon < \epsilon_c$ to globally chaotic dynamics for $\epsilon > \epsilon_c$. At this parameter and beyond, invariant spanning curves are no longer present. Depending on the initial conditions, chaos can diffuse unbounded in the phase space.

The determinant of the Jacobian matrix is $\det J = (1 - \gamma)$, and for $\gamma \neq 0$, Liouville's theorem is violated, directing to the existence of attractors in the phase space. For large enough ϵ, typically $\epsilon > 10$ sinks are not observed in the phase space hence leading the dynamics to have chaotic attractors in the limit of small values of γ. At such a limit, one faces a transition from limited ($\gamma \neq 0$) to unlimited ($\gamma = 0$) diffusion for the variable I, which is the transition we consider here using a phenomenological procedure.

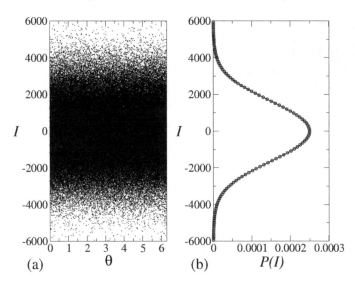

(a)   (b)

Fig. 7.1 **a** Plot of the phase space for a standard dissipative mapping considering the parameters $\epsilon = 100$ and $\gamma = 10^{-3}$. **b** The normalized probability distribution for the chaotic attractor shown in **a**

Figure 7.1a shows a plot of the phase space for the standard mapping considering the parameters $\epsilon = 100$ and $\gamma = 10^{-3}$. The figure shows a cloud of points with a large density near the origin. It then fades off as soon as the action is far from its origin. Figure 7.1b shows the probability density $P(I)$ exhibiting high concentration near the origin with a typical Gaussian shape.

The natural observable to characterize the diffusion is the average squared action $I_{rms}(n) = \sqrt{\frac{1}{M} \sum_{i=1}^{M} I_i^2}$ where M corresponds to an ensemble of distinct initial conditions along the chaotic attractor. The behavior of I_{rms} is exhibited in Fig. 7.2a.

Based on the behavior of the curves illustrated in Fig. 7.2a, the following scaling hypotheses can be proposed:

- Assuming $I_0 \cong 0$ and considering short time, normally $n \ll n_x$

$$I_{rms} \propto (n\epsilon^2)^\beta, \tag{7.2}$$

with β symbolizing the acceleration exponent;
- For large enough time, say $n \gg n_x$ then

$$I_{sat} \propto \epsilon^{\alpha_1}(1 - \gamma)^{\alpha_2}, \tag{7.3}$$

with both α_i, $i = 1, 2$ corresponding to the saturation exponents;
- Eventually, the crossover iteration number, marking the transformation from the regime of growth with $I_0 \cong 0$ to the saturation, is provided by

$$n_x \propto \epsilon^{z_1}(1 - \gamma)^{z_2}, \tag{7.4}$$

with z_1 and z_2 denoting the changeover exponents.

Fig. 7.2 a Plot of the phase space for the standard dissipative mapping assuming the parameters $\epsilon = 100$ and $\gamma = 10^{-3}$. **b** The normalized probability distribution for the chaotic attractor shown in **a**. The inset of **b** shows an exponential decay to the attractor

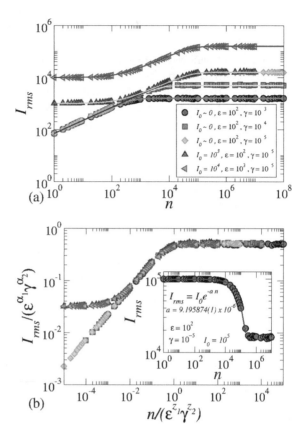

Using similar arguments as those discussed in Chap. 3, we assume the behavior of I_{rms} can be described by a homogeneous and generalized function of their variables of the type

$$I_{rms}(n\epsilon^2, \epsilon, (1-\gamma)) = \ell I_{rms}(\ell^a n\epsilon^2, \ell^b \epsilon, \ell^c (1-\gamma)), \qquad (7.5)$$

where a, b and c are named as characteristic exponents. After executing analogous procedures as made in Chap. 3, the two scaling laws are obtained

$$z_1 = \frac{\alpha_1}{\beta} - 2, \qquad (7.6)$$

$$z_2 = \frac{\alpha_2}{\beta}. \qquad (7.7)$$

A power law fitting for the growth regime beginning with $I_0 \cong 0$ provides $\beta = 1/2$. Therefore, the chaotic dynamics exhibits features analogous to the dynamics of random walk particles. The saturation regimes give $\alpha_1 = 1$ and $\alpha_2 = -1/2$ while the changeover exponents are obtained as $z_1 = 0$ and $z_2 = -1$.

7.2　Analytical Description of the Diffusion

Two ranges of parameters drive the investigation and lead to a scaling invariance. The first considers γ positive and small, typically $\gamma \in [10^{-5}, 10^{-2}]$. The second assumes $\epsilon > 10$, which drives the system to high nonlinearities and, nonetheless, to the absence of sinks in the phase space. As shown in Fig. 7.1a the chaotic attractor observed for the parameters $\epsilon = 10$ and $\gamma = 10^{-3}$ has positive Lyapunov exponent measured as $\lambda = 3.9120(1)$.

If we start the dynamics from an initial action $I_0 \cong 0$, the particle is permitted to diffuse along the chaotic attractor. In this section, the diffusion is described using the diffusion equation, which gives probability density to observe a specific action I at a given time n, i.e. $P(I, n)$. The diffusion equation is written as

$$\frac{\partial P(I, n)}{\partial n} = D \frac{\partial^2 P(I, n)}{\partial I^2}, \tag{7.8}$$

where the diffusion coefficient D is obtained from the first equation of the mapping by using $D = \frac{\overline{I^2_{n+1} - I^2_n}}{2}$. A straightforward calculation taking into account statistical independence between I_n and θ_n at the chaotic domain gives

$$D(\gamma, \epsilon, n) = \frac{\gamma(\gamma - 2)}{2}\overline{I^2}_n + \frac{\epsilon^2}{4}. \tag{7.9}$$

The expression of $\overline{I^2}_n$ is also obtained from the first equation of the mapping, assuming that $\overline{I^2}_{n+1} - \overline{I^2}_n = \frac{\overline{I^2}_{n+1} - \overline{I^2}_n}{(n+1)-n} \cong \frac{d\overline{I^2}}{dn} = \gamma(\gamma - 2)\overline{I^2} + \frac{\epsilon^2}{2}$, giving as the solution

$$\overline{I^2}(n) = \frac{\epsilon^2}{2\gamma(2 - \gamma)} + \left(I_0^2 + \frac{\epsilon^2}{2\gamma(\gamma - 2)}\right)e^{-\gamma(2-\gamma)n}. \tag{7.10}$$

To compare with the simulations, Eq. (7.10) has to be averaged over the orbit, hence

$$<\overline{I^2}(n)> = \frac{1}{n+1}\sum_{i=0}^{n}\overline{I^2}(i) = \frac{\gamma(\gamma - 2)}{2(n + 1)}$$

$$\times \left[I_0^2 + \frac{\epsilon^2}{2\gamma(\gamma - 2)}\left(\frac{1 - e^{-(n+1)\gamma(2-\gamma)}}{1 - e^{-\gamma(2-\gamma)}}\right)\right]. \tag{7.11}$$

An unique solution of Eq. (7.8) is obtained setting the following boundary conditions $\lim_{I \to \pm\infty} P(I) = 0$ with the initial condition $P(I, 0) = \delta(I - I_0)$. This specific choice characterizes all particles left with the same initial action but with M different initial phases $\theta \in [0, 2\pi]$.

The diffusion coefficient D depends on n. However, its variation is slow from the instant n to $n + 1$. This characteristic allows us to assume it as a constant, a procedure that facilitates obtaining the solution of the diffusion equation. However, as soon as the solution is encountered, the expression of D from Eq. (7.9) is incorporated into the solution. The agreement of this theoretical procedure with the numerical simulations is excellent.[1]

The technique used to solve Eq. (7.8) is the Fourier transform, present in many textbooks. Given the probability is normalized, $\int_{-\infty}^{\infty} P(I, n)dI = 1$, the following function can be used

$$R(k, n) = \mathcal{F}\{P(I, n)\} = \frac{1}{\sqrt{2\pi}} \int_{-\infty}^{\infty} P(I, n)e^{ikI} dI. \tag{7.12}$$

Differentiating $R(k, n)$ concerning n and from the property that $\mathcal{F}\left\{\frac{\partial^2 P}{\partial I^2}\right\} = -k^2 R(k, n)$ the following equation has to be solved $\frac{dR}{dn}(k, n) = -Dk^2 R(k, n)$, giving

$$R(k, n) = R(k, 0)e^{-Dk^2 n}. \tag{7.13}$$

Using the initial condition $P(I, 0) = \delta(I - I_0)$, then $R(k, 0) = \mathcal{F}\{\delta(I - I_0)\} = \frac{1}{\sqrt{2\pi}}e^{ikI_0}$. Inverting the expression of $R(k, n)$ gives

$$P(I, n) = \frac{1}{\sqrt{2\pi}} \int_{-\infty}^{\infty} R(k, n)e^{-ikI} dk$$

$$= \frac{1}{\sqrt{4\pi Dn}}e^{-\frac{(I-I_0)^2}{4Dn}}. \tag{7.14}$$

Equation (7.14) fulfills both the boundary and initial conditions as well as the diffusion equation (7.8). It is also normalized by construction. The observable of interest to characterize the diffusion is $\overline{I^2}(n) = \int_{-\infty}^{\infty} I^2 P(I, n)dI$, which leads to $\overline{I^2}(n) = \sqrt{2D(n)n + I_0^2}$. Using $D(n)$ obtained from Eq. (7.9), yields the expression of $I_{rms}(n)$ to be written as

$$I_{rms}(n) = \sqrt{I_0^2 + \frac{n\gamma(\gamma - 2)}{n + 1}\left[I_0^2 + \frac{\epsilon^2}{2\gamma(\gamma - 2)}\right]\left[\frac{1 - e^{-(n+1)\gamma(2-\gamma)}}{1 - e^{-\gamma(2-\gamma)}}\right]}. \tag{7.15}$$

Equation (7.15) allows to discuss specific limits of n and its consequences.

The first limit is $n = 0$, which leads to $I_{rms}(0) = I_0$, in agreement with the initial condition.

The second limit is $n \to \infty$. Such limit outcomes

$$I_{rms} = \sqrt{I_0^2 + \gamma(\gamma - 2)\left[I_0^2 + \frac{\epsilon^2}{2\gamma(\gamma - 2)}\frac{1}{1 - e^{-\gamma(2-\gamma)}}\right]}, \tag{7.16}$$

---

[1] We emphasize using this approximation, even with a good agreement between the theoretical with the numerical, the analytical expression obtained for P no longer satisfies the diffusion equation.

and that when developing in Taylor series up to first order, the term $1 - e^{-\gamma(2-\gamma)} \cong \gamma(2 - \gamma)$ gives

$$I_{rms} = \frac{1}{\sqrt{2(2-\gamma)}} \epsilon \gamma^{-1/2}. \tag{7.17}$$

This result shows analytically the two critical exponents α_1 and α_2. In the stationary state, the one obtained for $n \to \infty$, is given by $I_{rms} \propto \epsilon^{\alpha_1} \gamma^{\alpha_2}$. An immediate comparison of this scaling hypothesis with Eq. (7.17) leads to remarkable results of $\alpha_1 = 1$ and $\alpha_2 = -\frac{1}{2}$, in very well agreement with the phenomenological approach. This conclusion can be obtained from the equations of the mapping imposing $\overline{I^2}_{n+1} = \overline{I^2}_n = \overline{I^2}_{sat}$, yielding $I_{sat} = \frac{1}{\sqrt{2(2-\gamma)}} \epsilon \gamma^{-1/2}$.

The limit of small n is the third we consider. Assuming that the initial action $I_0 \cong 0$, consequently negligible as compared to ϵ and accomplishing a Taylor expansion on the exponential of the numerator from Eq. (7.16) provides $I_{rms}(n) \cong \sqrt{\frac{\epsilon^2}{2}n}$. This result demonstrates that for short n, an ensemble of particles diffuses along the chaotic attractor analogously as a random walk motion, therefore with diffusion exponent $\beta = 1/2$, i.e., normal diffusion. As discussed earlier, the limit of small n is $I_{rms}(n) \propto (n\epsilon^2)^{\beta}$, with $\beta = 1/2$ in well agreement with the theoretical prediction discussed above.

A fourth limit considers an intermediate n but non-negligible I_0 such as $0 < I_0 < I_{sat}$. At such windows of I_0 and n, an additional crossover is observed when $n'_x \cong 2\frac{I_0^2}{\epsilon^2}$. The existence of the plateau resembles the symmetry of probability. At that domain, half of the ensemble increases the action I, and the other half decreases. However, the decrease is interrupted when the particles achieve the lowest action permitted, breaking the symmetry and leading to the addition crossover.

A fifth limit is in the case of $I_0 \cong 0$. It leads to a growth in I_{rms} for short n followed by a crossover and a bend towards the saturation regime. Such a characteristic crossover is given by $n_x \cong \frac{1}{2-\gamma}\gamma^{-1}$. As discussed, the scaling approach assumes that $n_x \propto \epsilon^{z_1}\gamma^{z_2}$ and that $z_1 = 0$ and $z_2 = -1$, as discussed in the earlier section.

The last regime of interest is considered when $I_0 \gg \frac{\epsilon^2}{2\gamma(2-\gamma)}$. At this limit, Eq. (7.15) is rewritten as

$$I_{rms}(n) = \sqrt{I_0^2 e^{-(n+1)\gamma(2-\gamma)} + \epsilon^2 \frac{(1 - e^{-(n+1)\gamma(2-\gamma)})}{2\gamma(2-\gamma)}}. \tag{7.18}$$

The leading term for small n is $I_{rms}(n) = I_0 e^{-(n+1)\frac{\gamma(2-\gamma)}{2}}$ while the stationary state is obtained at the limit of $\lim_{n\to\infty} I_{rms} = \frac{\epsilon}{\sqrt{2(2-\gamma)}}\gamma^{-1/2}$, in well agreement with the previous results.

Figure 7.2a shows a plot of I_{rms} versus n for different control parameters and initial conditions, as labeled in the figure. Filled symbols correspond to the numerical simulation obtained directly from the iteration of the dynamical equations of the mapping considering an ensemble of $M = 10^3$ different initial particles. All the particles start with the same action I_0, as shown in Fig. 7.2a and different initial phases $\phi_0 \in [0, 2\pi]$. The analytical result from Eq. (7.15) is plotted as a continuous line. The overlap of the curves is excellent. Figure 7.2b shows the overlap of the curves plotted in (a) onto a single and hence universal curve. The scaling transformations are: (i) $I_{rms} \rightarrow I_{rms}/(\epsilon^{\alpha_1}\gamma^{\alpha_2})$; (ii) $n \rightarrow n/(\epsilon^{z_1}\gamma^{z_2})$. The inset of Fig. 7.2b shows the exponential decay as predicted by Eq. (7.18). The control parameters used in the inset were $\epsilon = 10^2$ and $\gamma = 10^{-5}$ and with the initial action $I_0 = 10^5$. The slope of the exponential decay obtained numerically is $a = 9.195874(1) \times 10^{-6}$, which is close to $\gamma(2 - \gamma)/2 \cong 9.99995 \times 10^{-6}$.

7.3 Summary

The diffusion equation describes a scaling invariance in a dissipative standard mapping. A set of critical exponents was obtained analytically, corroborating a substantial and general interest in the procedure.

Chapter 8
Characterization of a Transition from Limited to Unlimited Diffusion

Abstract A transition from limited to unlimited diffusion is described for a dissipative standard mapping. A two-dimensional mapping describes the system's dynamics and has two control parameters. One controls the intensity of the nonlinearity, while the other controls the dissipation. The parameter responsible for the transition is the dissipation, while the parameter driving the nonlinearity gives the elementary excitation of the dynamics. An order parameter is identified and goes continuously to zero at the transition. Moreover, its susceptibility diverges at the same limit. These elements give evidence the transition is characterized as a continuous phase transition.

8.1 Pieces of Evidence of a Phase Transition

As discussed in the previous chapter, a dissipative version of the standard mapping leads to the suppression of the unlimited chaotic diffusion. It happens because the determinant of the Jacobian matrix is smaller than one, violating Liouville's theorem. As an immediate consequence, a set of points where initial conditions converge are observed in the phase space. Due to the exponential spread in time of closely initial conditions, the attractors observed for a specific range of control parameters are chaotic. There are two control parameters describing the mapping

$$\begin{cases} I_{n+1} = (1 - \gamma)I_n + \epsilon \sin(\theta_n), \\ \theta_{n+1} = (\theta_n + I_{n+1}) \bmod(2\pi), \end{cases} \tag{8.1}$$

where $\gamma \in [0, 1]$ is the dissipative parameter, and ϵ corresponds to the nonlinearity strength. The parameter ϵ controls two transitions well known in the literature. For the conservative case of $\gamma = 0$, the parameter $\epsilon = 0$ makes the system integrable. Among the energy, the action I is also a constant of motion. The phase space is foliated, and no chaos is observed. For $\epsilon \neq 0$, the phase space turns mixed and periodic islands, invariant curves, and even chaos can all be observed. A specific transition happens at $\epsilon_c = 0.9716\ldots$, where the invariant spanning curves separating different portions of the phase space are destroyed. This destruction allows chaos to diffuse unbounded in the phase space; therefore, the transition is from locally chaotic, for $\epsilon < \epsilon_c$, to globally chaotic dynamics when $\epsilon > \epsilon_c$. Our interest lies in a

45

range of control parameters leading to unlimited diffusion in the phase space. Hence we considered $\epsilon \geq 10$, defining large nonlinearity. The diffusion is so far suppressed by the presence of the parameter $\gamma \neq 0$, which corresponds to a dissipative term. Indeed, the limit range of interest is $\gamma \in [0, 1]$ and always $\gamma \approx 0^+$.

The observable used to investigate the chaotic diffusion is the root square of the squared action, as defined in the previous chapter as I_{rms}. The discussion during the last chapter characterized the behavior of I_{rms} via a homogeneous and generalized function, implying a scaling invariance. The scaling invariance was confirmed via the existence of two scaling laws with specific critical exponents that led to universal curves while plotted in scaling variables. This scaling invariance is evidence of a possible phase transition faced by the chaotic diffusion.

In this chapter, we answer a set of four questions giving pieces of evidence the phase transition experienced by chaotic diffusion is of second order, consequently a continuous phase transition. The specific questions are (1) Identify the broken symmetry; (2) Define the order parameter; (3) Discuss the elementary excitation and; and (4) Discuss the topological defects which impact the transport of particles.

8.2   Break of Symmetry

The transition in light is the suppression of unlimited diffusion, hence a transition from limited to unlimited diffusion. We assume large values of the parameter ϵ, while the parameter that suppresses the diffusion is γ. Figure 8.1 shows the behavior for I_{rms} versus n for different control parameters and initial conditions. We noticed the curves saturate for a large enough time. Starting the dynamics with low initial action, typically $I_0 \cong 0$ the curves start to grow with n in a power law style with a slope of $(n\epsilon^2)^\beta$ with $\beta = 1/2$. In the absence of dissipation, the curve is symmetric and grows unbounded. However, when dissipation is introduced considering $\gamma > 0$, the violation of Liouville's theorem implies the existence of attractors. Since such an attractor is far from infinity, the unbounded diffusion is no longer observed. Hence,

Fig. 8.1 Plot of I_{rms} versus n for different control parameters and initial conditions, as labeled in the figure

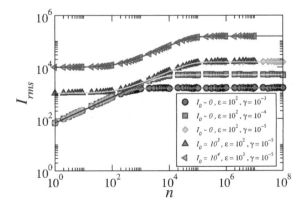

the behavior discussed in the previous chapter with the scaling invariance is happening. The curves of I_{rms} then saturate at a place that depends on the control parameters, especially from $I_{sat} \propto \epsilon^{\alpha_1} \gamma^{\alpha_2}$ where $\alpha_1 = 1$ and $\alpha_2 = -1/2$. The unlimited diffusion is then suppressed, destroying the symmetry of the unlimited growth, marking a break of symmetry.

Considering the equation of the mapping leading to the diffusion, we have $I_{n+1} = (1 - \gamma)I_n + \epsilon \sin(\theta_n)$. In the conservative case, $\gamma = 0$. When the dynamics grows unbounded, $I \gg \epsilon$ at a point where the nonlinear term can be neglected. Nonetheless, when unbounded growth is observed in the high-energy domain, $I_{n+1} = I_n$ is symmetric. When dissipation is present, such symmetry is broken since $I_{n+1} = (1 - \gamma)I_n$.

8.3 Order Parameter

The phenomenon discussed is the suppression of the unlimited chaotic diffusion for chaotic orbits due to dissipation in the dynamics. The break of symmetry is linked to the destruction of unbounded growth. In the presence of dissipation and considering long enough time, the curves of I_{rms} saturate at a constant plateau denoted as I_{sat}. The inverse of I_{sat} is a finite number when the dynamics is dissipative and is zero in the non-dissipative case. An order parameter is given by

$$\sigma = \frac{1}{I_{sat}},$$
$$\propto \gamma^{1/2}. \tag{8.2}$$

In the limit of $\gamma \to 0$, the order parameter σ approaches zero continuously. The susceptibility of the order parameter is

$$\xi = \frac{\partial \sigma}{\partial \gamma},$$
$$\propto \gamma^{-1/2}, \tag{8.3}$$

and corresponds to the response of the order parameter due to a variation in the intensity of the dissipation. The susceptibility diverges as soon as the control parameter γ approaches zero. Therefore, the order parameter goes continuously to zero at the same time its susceptibility diverges in the limit of parameter $\gamma \to 0$, both characteristics of a continuous phase transition.

8.4 Topological Defects

As we discussed in an earlier chapter, in the conservative case, the topological defects were associated with periodic islands, hence elliptical fixed points. Trapping dynamics leads to local confinement for a specific time interval when an orbit passes sufficiently close to a regular region. The phenomenon is called stickiness and affects the distribution probability along the chaotic sea hence modifying the diffusion locally. Near the trapping, the diffusion coefficient is no longer a constant as it is assumed along the chaotic sea far from the regular domains.

In the dissipative case, periodic islands are not observed. Objects in the phase space that would lead to modifications in the probability distribution in dissipative dynamics are sinks. They correspond to attracting fixed points of period one or larger periodic orbits. However, for the range of control parameters considered, particularly in the domain of large nonlinearity, sinks are not observed in the dynamics. Therefore, an initial condition in the chaotic region would diffuse without the chance of getting trapped by an attractive fixed point or periodic orbit. Moreover, the diffusion coefficient D assumes the following form:

$$D(\gamma, \epsilon, n) = \frac{\gamma(\gamma - 2)}{2}\overline{I^2}_n + \frac{\epsilon^2}{4}, \tag{8.4}$$

where $\overline{I^2}_n$ is given by

$$\overline{I^2}(n) = \frac{\epsilon^2}{2\gamma(2 - \gamma)} + \left(I_0^2 + \frac{\epsilon^2}{2\gamma(\gamma - 2)}\right)e^{-\gamma(2-\gamma)n}. \tag{8.5}$$

In the limit of $n \to \infty$, the diffusion coefficient $D \to 0$ leads the dynamics to reach the stationary distribution.

8.5 Elementary Excitations

The mapping considered is written in terms of two control parameters. One of them is dissipation $\gamma \in [0, 1]$ and the other one is ϵ that defines the length the particle can step from iteration n to $n + 1$. The parameter ϵ controls the intensity of the nonlinearity. Although ϵ is not related to the transition from bounded to unbounded diffusion,[1] it plays an important role in controlling the nonlinearity. Hence, it is responsible for producing the diffusion in the phase space, defining the elementary excitation of the dynamics.

8.6 Summary

In summary, we investigated the elements used to identify, in a dissipative standard mapping, a second-order phase transition. The scaling for chaotic diffusion happens due to the dissipation, producing chaotic attractors in the phase space. An order

---

[1] Indeed it is not connected to a criticality properly.

parameter was proposed as $\sigma = 1/I_{sat}$ that goes continuously to zero in the limit of $\gamma \to 0^+$. The susceptibility χ diverges in the same range. These two results are pieces of evidence of continuous phase transitions. The nonlinear function produces the elementary excitation, leading the dynamics to behave as random walk dynamics. The topological defects affecting the chaotic diffusion are associated with the attracting fixed points, not observed for the specific range of control parameters considered in the chapter. Therefore, the discussion presented here allows us to conclude that the phase transition from limited to unlimited chaotic diffusion in a dissipative version of the standard mapping is analogous to a second-order phase transition.

Chapter 9
Billiards with Moving Boundary

Abstract We discuss in this chapter a time-dependent billiard, a model with characteristics to produce unlimited diffusion in energy. We construct the equations of the mapping that describe the particles' dynamics, considering that the particle's velocity is given by the application of the momentum conservation law at each impact with the moving boundary. The unlimited diffusion is measured in the average speed of the particle leading to a phenomenon called Fermi acceleration.

9.1 A Billiard Dynamical System

Billiard is given by the dynamics of a classical particle confined to move inside a boundary with which it collides. The specular reflection guarantees that the angle the particle makes with a tangent line at the collision point after the collision is the same as before the collision. The particle's velocity changes since the boundary moves in time. Figure 9.1 shows a sketch of four collisions with the time-dependent boundary.

The position of the boundary is given in terms of polar coordinate $R = R_b(\theta, t)$. The dynamics of the particle is given by a four-dimensional nonlinear mapping written as $T(\theta_n, \alpha_n, V_n, t_n) = (\theta_{n+1}, \alpha_{n+1}, V_{n+1}, t_{n+1})$, where the variables are θ, denoting the angular position along the boundary where the impact happens; α, giving the angle the trajectory of the particle makes with a tangent line at the instant of the impact; V corresponds to the absolute value of the particle's velocity; t furnishes the instant of the impact. All variables are evaluated immediately after the collision n with the boundary.

The mapping is constructed starting the dynamics from an initial condition $(\theta_n, \alpha_n, V_n, t_n)$. The position of the particle is $X(\theta_n) = R(\theta_n, t_n)\cos(\theta_n)$ and $Y(\theta_n) = R(\theta_n, t_n)\sin(\theta_n)$ with the velocity written as

$$\vec{V}_n = |\vec{V}_n|[\cos(\phi_n + \alpha_n)\hat{i} + \sin(\phi_n + \alpha_n)\hat{j}], \tag{9.1}$$

Fig. 9.1 Plot of four
collisions of a particle with a
time-dependent boundary.
The position of the boundary
is drawn at the instant of the
impact

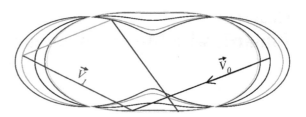

where $\hat{i}$ and $\hat{j}$ represent the unity vectors concerning the axes X and Y. The angle
ϕ_n is given by

$$\phi_n = \operatorname{atan}\left[\frac{Y'(\theta_n, t_n)}{X'(\theta_n, t_n)}\right], \tag{9.2}$$

where $X'(\theta_n, t_n) = dX(\theta_n, t_n)/d\theta_n$ and $Y'(\theta_n, t_n) = dY(\theta_n, t_n)/d\theta_n$.
    The position of the particle for $t \geq t_n$ is given by

$$X_\rho(t) = X(\theta_n) + |\vec{V}_n|\cos(\phi_n + \alpha_n)(t - t_n), \tag{9.3}$$

$$Y_\rho(t) = Y(\theta_n) + |\vec{V}_n|\sin(\phi_n + \alpha_n)(t - t_n). \tag{9.4}$$

The index ρ identifies the particle coordinates while b represents the boundary coordinates. The distance of the particle measured concerning the origin of a rectangular
coordinate system is written as $R_\rho(t) = \sqrt{X_\rho^2(t) + Y_\rho^2(t)}$. The angular position of
the particle at the next collision with the boundary is θ_{n+1} which is obtained from
the solution of

$$R_\rho(\theta_{n+1}, t_{n+1}) = R_b(\theta_{n+1}, t_{n+1}). \tag{9.5}$$

Among the angular position θ_{n+1}, the instant of the collision is given by

$$t_{n+1} = t_n + \frac{\sqrt{[\Delta X_\rho]^2 + [\Delta Y_\rho]^2}}{|\vec{V}_n|}, \tag{9.6}$$

where $\Delta X_\rho = X_\rho(\theta_{n+1}) - X_\rho(\theta_n)$ is $\Delta Y_\rho = Y_\rho(\theta_{n+1}) - Y_\rho(\theta_n)$.
    Since the referential frame of the boundary is accelerated, the velocity of the
particle follows the reflection laws

$$\vec{V}'_{n+1} \cdot \vec{T}_{n+1} = \vec{V}'_n \cdot \vec{T}_{n+1}, \tag{9.7}$$

$$\vec{V}'_{n+1} \cdot \vec{N}_{n+1} = -\vec{V}'_n \cdot \vec{N}_{n+1}, \tag{9.8}$$

where the primes correspond to the velocity of the particle measured in the referential
frame of the moving boundary. At θ_{n+1}, the unit vectors are written as

$$\vec{T}_{n+1} = \cos(\phi_{n+1})\hat{i} + \sin(\phi_{n+1})\hat{j}, \tag{9.9}$$

$$\vec{N}_{n+1} = -\sin(\phi_{n+1})\hat{i} + \cos(\phi_{n+1})\hat{j}, \tag{9.10}$$

leading to

$$\vec{V}_{n+1} \cdot \vec{T}_{n+1} = |\vec{V}_n|[\cos(\alpha_n + \phi_n) \cos(\phi_{n+1})]$$
$$+ |\vec{V}_n|[\sin(\alpha_n + \phi_n) \sin(\phi_{n+1})], \tag{9.11}$$

$$\vec{V}_{n+1} \cdot \vec{N}_{n+1} = -|\vec{V}_n|[-\cos(\alpha_n + \phi_n) \sin(\phi_{n+1})]$$
$$- |\vec{V}_n|[\sin(\phi_n + \alpha_n) \cos(\phi_{n+1})]$$
$$+ 2\vec{V}_b(t_{n+1}) \cdot \vec{N}_{n+1}, \tag{9.12}$$

where $\vec{V}_b$ is the velocity of the boundary given by

$$\vec{V}_b(t_{n+1}) = \frac{dR_b(t_{n+1})}{dt_{n+1}}[\cos(\theta_{n+1})\hat{i} + \sin(\theta_{n+1})\hat{j}], \tag{9.13}$$

and $\frac{dR_b(t_{n+1})}{dt_{n+1}}$ gives the velocity of the boundary at the impact $n + 1$.

Finally, $V_{n+1} = \sqrt{(\vec{V}_{n+1} \cdot \vec{T}_{n+1})^2 + (\vec{V}_{n+1} \cdot \vec{N}_{n+1})^2}$. The angle α_{n+1} is given by

$$\alpha_{n+1} = \operatorname{atan}\left[\frac{\vec{V}_{n+1} \cdot \vec{N}_{n+1}}{\vec{V}_{n+1} \cdot \vec{T}_{n+1}}\right]. \tag{9.14}$$

Given the dynamical variables of the billiard, numerical simulations can now be discussed.

9.2 Unlimited Diffusion of the Velocity for the Oval Billiard

Let us consider the application of the billiard formalism to an oval billiard. The radius of the billiard in polar coordinates is written as

$$R(\theta, p, \epsilon, \eta, t) = 1 + \epsilon(1 + \eta \cos(t)) \cos(p\theta), \tag{9.15}$$

where η is the amplitude of the time-dependent perturbation. Each particle's dynamics is made using the discrete mapping obtained in the previous section. The equation determines the average velocity is $\overline{V} = \frac{1}{M} \sum_{i=1}^{M} \frac{1}{n} \sum_{j=1}^{n} V_{i,j}$ where M gives the number of different initial conditions for an ensemble of initial conditions and n denotes the number of collisions each particle has with the boundary. Figure 9.2 shows a plot of the average velocity for the parameters $\epsilon = 0.08$, $p = 3$, $\eta = 0.5$ and different values for the initial conditions, as shown in the figure. When the initial velocity V_0 is sufficiently small, the average velocity $\overline{V}$ grows with a power law in n. As soon as the initial velocity increases, the curves of average velocity exhibit a plateau for a short time until they reach a crossover and turn towards a regime of growth represented by a power law. The behavior shown in Fig. 9.2 can be characterized by using the following scaling hypotheses: (i) for small enough initial velocities, naturally, from the order of the maximum velocity of the moving boundary, it is

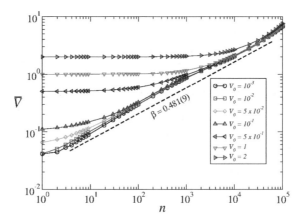

Fig. 9.2 Plot of the average velocity *versus* n for the control parameters: $\epsilon = 0.08$, $p = 3$, and $\eta = 0.5$. The initial velocities are shown in the figure

observed that $\overline{V} \propto n^{\beta}$; (ii) when the initial velocity is not small enough, a plateau is observed attending the conditions $\overline{V}_{plat.} \propto V_0^{\alpha}$, for $n \ll n_x$; and (iii) finally, the crossover number that marks the conversion from the constant plateau to the regime of growth is given by $n_x \propto V_0^z$. Here, α, β, and z are critical exponents. These three scaling hypotheses can be associated with a homogeneous and generalized function, as expressed in the previous chapters leading to the following scaling law $z = \alpha/\beta$. The exponent α is easy to be obtained. Since the curves stay constant for a large range of n, we conclude $\alpha = 1$. The acceleration exponent is obtained from a power law fitting and gives $\beta = 0.481(9)$. We obtain $z = 2.07(3)$ by using the scaling law.

From the knowledge of the critical exponents, the curves of the average velocity shown in Fig. 9.2 were obtained for different initial speeds that can be overlaid onto a single and, therefore, universal plot after the following scaling transformations: $\overline{V} \rightarrow \overline{V}/V_0^{\alpha}$ and $n \rightarrow n/V_0^z$. Figure 9.3 shows the overlap of the curves plotted in Fig. 9.2 after the scaling transformations.

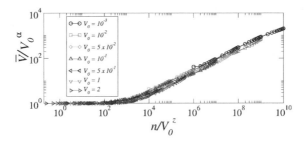

Fig. 9.3 Plot of the curves shown in Fig. 9.2 onto a single and universal curve after the following scaling transformations: $\overline{V} \rightarrow \overline{V}/V_0^{\alpha}$ and $n \rightarrow n/V_0^z$. The control parameters used are: $\epsilon = 0.08$, $p = 3$, and $\eta = 0.5$. The initial velocities are shown in the figure

The unlimited energy growth denoted by the growth of the curves corresponds to the unbounded diffusion for the average velocity, an observable similar to the one discussed previously for the standard model. This unlimited energy growth is not robust at a point where a tiny dissipation destroys the unbounded growth. This suppression also exhibits characteristics of a continuous phase transition, a subject discussed in the next chapter.

9.3 Summary

We discussed in this chapter some dynamic properties for a time-dependent oval billiard. We showed the Fermi acceleration is observed for the oval billiard. Depending on the initial conditions and control parameters, the velocity growth is described by using scaling hypotheses leading to a set of three critical exponents related to each other in a scaling law.

Chapter 10
Suppression of Fermi Acceleration in Oval Billiard

Abstract We discuss in this chapter some results concerning the suppression of unlimited energy diffusion, which is produced by introducing inelastic collisions of the particles with the boundary of the billiard. Two scaling laws emerge from a set of three scaling hypotheses. A scaling invariance is observed, giving pieces of evidence of a phase transition in the system.

10.1 Dissipation in a Billiard

We discuss in this section how to introduce dissipation in the equations describing the dynamics of a particle moving in a billiard with a moving boundary. The formalism is applied to an oval billiard. A fractional energy reduction at each impact of the particle with the border happens due to an inelastic collision with a restitution coefficient smaller than one.

The mapping construction is similar to that already discussed in Chap. 9, and the major difference is the reflection law. It is considered that the collisions of the particles with the boundary are inelastic, leading to a fractional energy loss upon each collision. It is assumed that only the normal component of the velocity is affected by the dissipation. In the instant of the collision, the reflection laws are

$$\vec{V}'_{n+1} \cdot \vec{T}_{n+1} = \vec{V}'_n \cdot \vec{T}_{n+1}, \tag{10.1}$$

$$\vec{V}'_{n+1} \cdot \vec{N}_{n+1} = -\gamma \vec{V}'_n \cdot \vec{N}_{n+1}, \tag{10.2}$$

where the unity normal and tangent vectors are given by

$$\vec{T}_{n+1} = \cos(\phi_{n+1})\hat{i} + \sin(\phi_{n+1})\hat{j}, \tag{10.3}$$

$$\vec{N}_{n+1} = -\sin(\phi_{n+1})\hat{i} + \cos(\phi_{n+1})\hat{j}, \tag{10.4}$$

with $\gamma \in [0, 1]$ corresponding to the restitution coefficient. Elastic collisions are recovered for $\gamma = 1$, while, for $\gamma < 1$, the particle has a fractional energy loss at each collision. The term $\vec{V}'$ corresponds to the velocity of the particle measured in the non-inertial referential frame where the collision happens. The components of the tangential and normal velocity at the impact $(n + 1)$ are given by

$$\vec{V}_{n+1} \cdot \vec{T}_{n+1} = \vec{V}_n \cdot \vec{T}_{n+1}, \tag{10.5}$$

$$\vec{V}_{n+1} \cdot \vec{N}_{n+1} = -\gamma \vec{V}_n \cdot \vec{N}_{n+1}$$
$$+ (1 + \gamma) \vec{V}_b(t_{n+1} + Z(n)) \cdot \vec{N}_{n+1}, \tag{10.6}$$

where $\vec{V}_b(t_{n+1} + 2\pi Z(n))$ corresponds to the velocity of the boundary, which is given by

$$\vec{V}_b(t_{n+1}) = \frac{dR(t)}{dt}\bigg|_{t_{n+1}} [\cos(\theta_{n+1})\hat{i} + \sin(\theta_{n+1})\hat{j}]. \tag{10.7}$$

The term $Z(n) \in [0, 1]$ is a random number introduced in the argument of the velocity of the boundary to model the stochasticity in the system and guarantee the absence of correlation between the other dynamical variables.

10.1.1 Long-Stand Dynamics

The dynamics for the prolonged time leading to the stationary state is obtained assuming the probability distribution for the velocity in the plane α versus θ is uniform. Doing an average for the quadratic velocity $|\vec{V}_{n+1}|$, we obtain

$$\overline{V^2}_{n+1} = \frac{\overline{V^2}_n}{2} + \frac{\gamma^2 \overline{V^2}_n}{2} + \frac{(1 + \gamma)^2 \eta^2 \varepsilon^2}{8}. \tag{10.8}$$

The average quadratic velocity is obtained assuming that $\overline{V_{n+1}^2} = \overline{V_n^2} = \overline{V^2}$, that leads to

$$\overline{V^2} = \frac{(1 + \gamma)\eta^2 \varepsilon^2}{4(1 - \gamma)}. \tag{10.9}$$

Extracting the square root from the average quadratic velocity $\overline{V} = \sqrt{\overline{V^2}}$ gives

$$\overline{V} = \frac{\eta\varepsilon}{2}\sqrt{(1 + \gamma)}(1 - \gamma)^{-1/2}. \tag{10.10}$$

We glimpse the exponent of the term $(1 - \gamma)$ is $-1/2$, while the exponent $(\eta\varepsilon)$ is 1 and either has an essential role in the scaling characterization.

10.1.2 Time-Dependent Regime

Iterating the dynamical equations of the mapping is a simple process. However, extracting analytical results from the equations involves some transformations.

One of them considered is transforming the equation of differences describing the particle's velocity into an ordinary differential equation with a simple solution. Transforming Eq. (10.8) into an ordinary differential equation gives

$$\overline{V^2}_{n+1} - \overline{V^2_n} = \frac{\overline{V^2}_{n+1} - \overline{V^2_n}}{(n+1) - n} \cong \frac{d\overline{V^2}}{dn}, \qquad (10.11)$$

leading to

$$\frac{d\overline{V^2}}{dn} = \frac{\overline{V^2}}{2}(\gamma^2 - 1) + \frac{(1+\gamma)^2\eta^2\varepsilon^2}{8}. \qquad (10.12)$$

Integrating and considering V_0 at $n = 0$ gives

$$\overline{V^2}(n) = \overline{V_0^2}e^{\frac{(\gamma^2-1)}{2}n} + \frac{(1+\gamma)}{4(1-\gamma)}\eta^2\varepsilon^2\left[1 - e^{\frac{(\gamma^2-1)}{2}n}\right]. \qquad (10.13)$$

The dynamics of $\overline{V}(n) = \sqrt{\overline{V^2}(n)}$ is described by

$$\overline{V}(n) = \sqrt{\overline{V_0^2}e^{\frac{(\gamma^2-1)}{2}n} + \frac{(1+\gamma)}{4(1-\gamma)}\eta^2\varepsilon^2\left[1 - e^{\frac{(\gamma^2-1)}{2}n}\right]}. \qquad (10.14)$$

Considering $V_0 \gg \frac{(1+\gamma)^{1/2}}{2}(1 - \gamma)^{-1/2}\eta\varepsilon$ gives an exponential decay for the velocity as

$$\overline{V}(n) \cong V_0 e^{\frac{(\gamma^2-1)}{4}n} \cong V_0 e^{\frac{(\gamma-1)}{2}n}. \qquad (10.15)$$

Considering now a small velocity, say $V_0 \cong 0$, the dominant expression for $\overline{V}(n)$ is

$$\overline{V}(n) = \frac{(1+\gamma)^{1/2}}{2}(1 - \gamma)^{-1/2}\eta\varepsilon\left[1 - e^{\frac{(\gamma^2-1)}{2}n}\right]^{1/2}. \qquad (10.16)$$

A Taylor expansion for Eq. (10.16) gives

$$\overline{V}(n) \sim \eta\varepsilon\sqrt{n}, \qquad (10.17)$$

hence $\beta = 1/2$.

10.1.3 Discussion of the Phenomenological Results

We examine, from now on, the behavior of the average quadratic velocity for a set of particles using numerical simulations for the limit of $\gamma \to 1$. Such a range of γ characterizes the dynamics near a transition from conservative to dissipative dynamics.

For $\gamma = 1$, the average velocity for the ensemble of particles grows unlimitedly. When $0 < \gamma < 1$, the energy growth is not unbounded, producing a stationary state. The numerical simulations were made considering an initial velocity of $V_0 = 10^{-3}$, using the following range of control parameters $\eta\varepsilon \in [0.002, 0.02]$ with both the initial angles $\alpha_0 \in [0, \pi]$, $\theta_0 \in [0, 2\pi]$ as well as the initial time $t_0 \in [0, 2\pi]$ randomly chosen. At each collision, a random number $Z(n) \in [0, 1]$ is chosen, and the velocity of the boundary is determined. Two types of averages are considered

$$< \overline{V} > (n) = \frac{1}{M} \sum_{i=1}^{M} \frac{1}{n+1} \sum_{j=0}^{n} V_{i,j} . \qquad (10.18)$$

The index i corresponds to a sample in an ensemble of $M = 2000$ different initial conditions. A plot of $< \overline{V} >$ versus n for different values of γ is shown in Fig. 10.1a.

Figure 10.1a shows two different behaviors. For small values of n, the average velocity grows with power law and bends towards a regime of saturation for large enough n. A characteristic crossover n_x gives the changeover from growth to saturation. The transformation $n \rightarrow n(\eta\varepsilon)^2$ overlaps all the curves for a short time before they converge to saturation at different positions. From the behavior observed in Fig. 10.1a, the following scaling hypotheses can be proposed: (i) For values of $n \ll n_x$, the regime of growth is described by $< \overline{V} > \propto [(\eta\varepsilon)^2 n]^\beta$, where β is the acceleration exponent; (ii) for $n \gg n_x$ then $< \overline{V}_{sat} > \propto (1 - \gamma)^{\alpha_1}(\eta\varepsilon)^{\alpha_2}$, where α_1 and α_2 are saturation exponents; and (iii) the crossover number n_x that marks the changeover from growth to the saturation is given by $n_x \propto (1 - \gamma)^{z_1}(\eta\varepsilon)^{z_2}$, where z_1 and z_2 are crossover exponents.

The three scaling hypotheses allow the behavior of $< \overline{V} >$ to be described by a homogeneous and generalized function of the type

$$< \overline{V} > [(\eta\varepsilon)^2 n, \eta\varepsilon, (1 - \gamma)] = l < \overline{V} > [l^a(\eta\varepsilon)^2 n, l^b \eta\varepsilon, l^d(1 - \gamma)], \qquad (10.19)$$

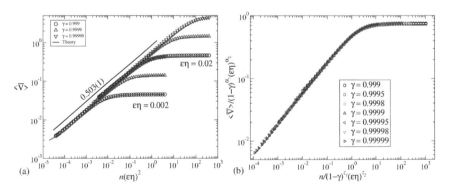

Fig. 10.1 **a** Plot of $< \overline{V} >$ versus n for different values of γ and two different combinations of $\eta\varepsilon$. **b** Overlap of the curves shown in **a** onto a single and universal plot after the application of the following scaling transformations: $n \rightarrow n/[(1 - \gamma)^{z_1}(\eta\varepsilon)^{z_2}]$ and $< \overline{V} > \rightarrow < \overline{V} > /[(1 - \gamma)^{\alpha_1}(\eta\varepsilon)^{\alpha_2}]$. The continuous lines give the theoretical results obtained from Eq. (10.23)

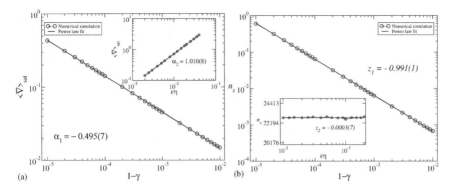

Fig. 10.2 Plot of: **a** $< \overline{V}_{sat} >$ and **b** n_x as a function of $(1 - \gamma)$. The inner plots show the behavior of $< \overline{V}_{sat} >$ and n_x for different values of $\varepsilon\eta$

where l is a scaling factor, a, b, and d are characteristic exponents that must be related to the critical exponents. Using a similar procedure as made in previous chapters, the following scaling laws are obtained

$$z_1 = \frac{\alpha_1}{\beta}, \qquad z_2 = \frac{\alpha_2}{\beta} - 2. \qquad (10.20)$$

All the five critical exponents can be obtained numerically. When fitting a power law to the velocity regime's growth, we have $\beta = 0.503(1) \simeq 1/2$. Considering now a fixed $\eta\varepsilon$ and varying γ, we obtain a power law fitting for $< \overline{V}_{sat} >$ versus $(1 - \gamma)$ that gives $\alpha_1 = -0.495(7) \cong -1/2$, as shown in Fig. 10.2a. A power law fitting for n_x versus $(1 - \gamma)$ gives that $z_1 = -0.991(1) \cong -1$. Finally, assuming $(1 - \gamma)$ as a constant and fitting a power law for $< \overline{V}_{sat} >$ versus $\eta\varepsilon$ gives $\alpha_2 = 1.010(8) \cong 1$, while a plot of n_x versus $\eta\varepsilon$ gives a fitting of $z_2 = -0.0003(7) \cong 0$. When the scaling laws are used to determine the critical exponents from Eq. (10.20), we detect an excellent agreement between the theoretical approach and the numerical results. Applying the conversions $n \rightarrow n/[(1 - \gamma)^{z_1}(\eta\varepsilon)^{z_2}]$ and $< \overline{V} > \rightarrow < \overline{V} > /[(1 - \gamma)^{\alpha_1}(\eta\varepsilon)^{\alpha_2}]$, all the curves shown in Fig. 10.1a overlap each other onto a single and hence universal plot as shown in Fig. 10.1b.

10.1.4 Time Average

As discussed in Eq. (10.13), the average quadratic velocity is obtained considering only the average over an ensemble of different initial conditions corresponding to an ensemble of non-interacting particles. However, the simulations made considered both ensemble and time averages. Hence, the average quadratic velocity is given by

$$< \overline{V^2}(n) >= \frac{1}{n+1} \sum_{i=0}^{n} \overline{V^2}(i). \qquad (10.21)$$

The summation emerging in the exponential converges since the arguments of the exponential function are negative. The convergence of the exponential is given by

$$\sum_{i=0}^{n} e^{(\frac{\gamma^2-1}{2})i} = \left[\frac{1 - e^{(\frac{\gamma^2-1}{2})(n+1)}}{1 - e^{\frac{\gamma^2-1}{2}}} \right],$$

(10.22)

and when taking the square root, we obtain $V_{rms}(n) = \sqrt{< \overline{V^2}(n) >}$, that leads to

$$V_{rms}(n) = \sqrt{\frac{(1+\gamma)\eta^2\varepsilon^2}{4(1-\gamma)} + \frac{1}{(n+1)} \left[V_0^2 - \frac{(1+\gamma)\eta^2\varepsilon^2}{4(1-\gamma)} \right] \left[\frac{1 - e^{(n+1)\frac{(\gamma^2-1)}{2}}}{1 - e^{\frac{(\gamma^2-1)}{2}}} \right]}.$$

(10.23)

A plot of the curve generated from Eq. (10.23) is represented by the continuous line in Fig. 10.1a.

Two crucial limits for Eq. (10.23) are:

1. $n = 0$, that leads to $V_{rms}(0) = V_0$;
2. Considering the limit of $n \to \infty$, we obtain

$$V_{rms}(n \to \infty) = \sqrt{\frac{(1+\gamma)\eta^2\varepsilon^2}{4(1-\gamma)}}.$$

(10.24)

From the Eq. (10.23), we can discuss the behavior of V_{rms} for small values of n. In the limit of $\gamma \approx 1$, the exponentials from Eq. (10.23) can be Taylor expanded. Due to the existence of the term $(n + 1)$ in the denominator of Eq. (10.23), a Taylor expansion must go until the second order. Nevertheless, the expansion in the denominator can be interrupted just in the first order. Doing Taylor expansions, the expression for $V_{rms}(n)$ in the limit of $V_0 \cong 0$ is given by

$$V_{rms}(n) \cong \frac{(1+\gamma)\eta\varepsilon}{4} \sqrt{(n+1)}.$$

(10.25)

When $n \gg 1$ implying that $\sqrt{(n+1)} \cong \sqrt{n}$ we obtain $V_{rms}(n) \cong \frac{(1+\gamma)\eta\varepsilon}{4} \sqrt{n}$.

10.1.5 Determination of the Critical Exponents

The five critical exponents that describe the scaling properties for the average quadratic velocity are β, α_i, and z_i with $i = 1, 2$. The exponents α_1 and α_2 are obtained for the regime of $n \to \infty$. From the Eq. (10.24), we obtain that $\alpha_1 = -1/2$ and $\alpha_2 = 1$. The exponent β is obtained from Eq. (10.25). When $n \gg 1$, we have that $\beta = 1/2$. Finally, the crossover n_x can be estimated from equaling Eqs. (10.25) and (10.24). An immediate algebra gives

$$n_x = \frac{4}{(1+\gamma)}(1-\gamma)^{-1}. \qquad (10.26)$$

We can conclude that $z_1 = -1$ and $z_2 = 0$.

10.1.6 Distribution of Velocities

We now discuss the shape of the velocity distribution for the dissipative dynamics in the presence of inelastic collisions. It is essential to notice that the particle's velocity has a lower limit given by the speed of the moving boundary, i.e., $V_l = -\eta\epsilon$. The upper limit depends on the control parameters, particularly the dissipation parameter. The lower limit of the velocity has a crucial role in the velocity distribution.

Suppose an ensemble of particles with different initial conditions α, θ but with the same initial velocity chosen beyond the lower limit and below the upper limit. The dynamics evolves such that for a short number of collisions of the particles with the boundary, part of the ensemble of particles raises velocity, while the other amount decreases speed. As shown in Fig. 10.3, the distribution is Gaussian for the initial velocity $V_0 = 0.2$ considering 10 collisions with the boundary. The control parameters used were $\epsilon\eta = 0.02$, $\gamma = 0.999$, and $p = 2$, although other

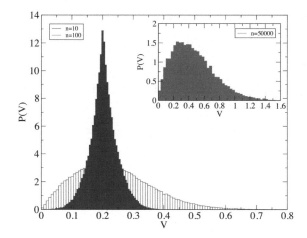

Fig. 10.3 Plot of the probability distribution for an ensemble of 10^5 particles in a dissipative and stochastic version of the oval billiard. Blue was obtained for 10 collisions with the boundary, while red was obtained for 100 collisions. The construction of the inset figure used 50,000 collisions. The initial velocity considered was $V_0 = 0.2$ and the control parameters used were $\epsilon\eta = 0.02$ and $\gamma = 0.999$ for $p = 2$

combinations produce similar results. We considered an ensemble of 2.5×10^6 different initial conditions. With the evolution of the dynamics, the Gaussian curve turns flat from both sides until it touches the lower limit of the velocity from the left. Such behavior can be seen from the bars for $n = 100$ collisions of the particles with the boundary. For the combination of control parameters used, from $n = 100$ collisions, the symmetry of the distribution is broken and is no longer Gaussian but rather exhibits a shape as shown in the inner part of Fig. 10.3. The distribution was obtained after $50,000$ collisions of the particles with the boundary. Although the distribution is no longer Gaussian resembling the Boltzmann distribution, it shows a peak and decays monotonically for higher values of velocity, warranting convergence of distribution momenta.

An observable giving information on the long-range correlation of the dynamics is the deviation around the average velocity. Nonetheless, it shows similar behavior as the one observed for the average velocity, as shown in Fig. 10.1. For a considerable enough time, the curves of average velocity reach the stationary estate. From that point and beyond, the curves of the deviation around the average velocity also saturate. They indicate a limited range of variation hence a finite fluctuation around an average velocity.

10.2 Summary

We discussed in this chapter some scaling properties present in the diffusion of energy for an ensemble of particles in a time-dependent billiard. We showed the behavior of the squared average velocity is described by a homogeneous and generalized function, leading to a set of two scaling laws given by

$$z_1 = \frac{\alpha_1}{\beta} \, , \, z_2 = \frac{\alpha_2}{\beta} - 2. \tag{10.27}$$

The numerical values for the critical exponents are $\beta = 1/2$, $\alpha_1 = -1/2$, $\alpha_2 = 1$, $z_1 = -1$, and $z_2 = 0$.

Chapter 11
Suppressing the Unlimited Energy Gain: Shreds of Evidence of a Phase Transition

Abstract Suppression of Fermi acceleration is described as a phase transition from limited to unlimited energy diffusion for a dissipative oval billiard. The dynamics of each particle in the billiard is made using a four-dimensional mapping with two relevant control parameters. One controls the intensity of the nonlinearity, and the other controls the amount of dissipation. An order parameter is identified with a parameter responsible for the transition and a parameter giving the elementary excitation of the dynamics.

11.1 Elements of a Phase Transition

As discussed in the previous chapter, the dissipation introduced in a time-dependent oval billiard was due to inelastic collisions. It implies that after each impact, the particle has a fractional loss of energy. The dynamics is made by using a four-dimensional mapping with two relevant control parameters. Unlimited energy growth is suppressed when the dissipation parameter is smaller than the unity.

In the conservative case, the observable leading to unlimited energy growth corresponds to the average velocity $\overline{V}(n)$, which grows in a power law fashion $V \propto n^\beta$. However, when dissipation is considered, unlimited growth no longer happens. The curves of $\overline{V}(n)$ in the growing domain bend towards a regime of saturation, leading to the dynamics of the stationary state. The curves are described by a homogeneous and generalized function leading to a scaling invariance. The scaling for the diffusion was confirmed via the existence of two scaling laws

$$z_1 = \frac{\alpha_1}{\beta} , \quad z_2 = \frac{\alpha_2}{\beta} - 2, \tag{11.1}$$

with specific critical exponents that led to universal curves while plotted in scaling variables. This scaling invariance is evidence of a possible phase transition in the curves leading to bounded from unbounded dynamics.

Our aim in the chapter is to move forward by giving responses to the following questions: Identify the broken symmetry. Define the order parameter. Discuss the elementary excitation. Discuss the topological defects which impact the transport of particles.

11.2   Break of Symmetry

The regime of unbounded growth of the curves $\overline{V}(n)$ is observed to be described as a power law n^{β} with $\beta > 0$ only in the conservative case. A power law marking the growing regime is symmetric and leads to Fermi acceleration. Nonetheless, this regime of growth is not robust when dissipation introduced via inelastic collisions is present. The dissipation yields in area contraction in the phase space, creating a limited attractor. The curves are described by a homogeneous and generalized function leading to a scaling invariance, which is clear evidence of a phase transition. As soon as the dissipation is current, the symmetric regime of growth is no longer observed, marking a symmetry break for the curves of $\overline{V}(n)$.

11.3   Order Parameter

The transition considers a bounded to unbounded energy growth, suppressing the Fermi acceleration. The dynamics considers two relevant control parameters and dictates the saturation for the average velocity. Considering $n \gg n_x$, the dynamics is described by $< \overline{V}_{sat} > \propto (1 - \gamma)^{\alpha_1} (\eta \varepsilon)^{\alpha_2}$, where $\alpha_1 = -1/2$ and $\alpha_2 = 1$. For a parameter, $\gamma \neq 1$, $< \overline{V}_{sat} >$ is a finite number and also its inverse. An order parameter for this transition is proposed as $\sigma = \frac{1}{<\overline{V}_{sat}>}$, that has a form

$$\sigma = \frac{1}{< \overline{V}_{sat} >},$$
$$\propto \sqrt{(1 - \gamma)}. \tag{11.2}$$

This parameter goes smoothly and continuously to zero in the limit of $\gamma \to 1^-$.

The susceptibility of the order parameter is defined as

$$\chi = \frac{\partial \sigma}{\partial \gamma},$$
$$\propto (1 - \gamma)^{-1/2}, \tag{11.3}$$

that diverges in the limit of $\gamma \to 1^-$. These two characteristics of σ and χ give evidence the transition is a second-order phase transition.

11.4   Topological Defects

In the dissipative case, and considering the regime of small dissipation produced by $\gamma \to 1$, pieces of evidence lead to the existence of a strange attractor in the phase space. A plot of V *versus* ϕ, where ϕ is the time modulated π suggests the presence of an attractor far from the infinity; hence, unbounded growth of velocity is not observed. Attracting fixed points, however, is overlooked. Their existence would lead to trapping points, i.e., points to where the trajectories would be drawn and, once reached, only leave the region after an external perturbation. Their absence confirms the absence of topological defects and, therefore, the absence of areas in the phase space perturbing the transport of particles in the phase space. Since the dynamics is dissipative, stickiness is not present in the phase space.

11.5 Elementary Excitations

The mapping considered is written in terms of two control parameters, one is $(1 - \gamma)$, and the other is $(\eta \varepsilon)$. The first one corresponds to the dissipation term, while the other relates to the step size of increase or decrease of the velocity. The dissipation parameter carries the criticality, while the second controls the nonlinearity strength. The parameter $(\eta \varepsilon)$ is responsible for the diffusion of the particles in the phase space. Hence, it corresponds to the own elementary excitation of the dynamics.

11.6 Summary

We investigated the elements used to identify a second-order phase transition in a time-dependent oval billiard. The scaling observed for the diffusion of energy is related to dissipation. An order parameter was proposed as $\sigma = 1/ < \overline{V}_{sat} >$ and goes continuously to zero in the limit of $\gamma \to 1^-$. The susceptibility χ diverges in the same limit. These two results are pieces of evidence of continuous phase transitions. The nonlinear function produces the elementary excitation, leading the dynamics to behave as random walk dynamics. The topological defects affecting the diffusion are associated with the attracting fixed points, not observed for the specific range of control parameters considered in the chapter. Therefore, the discussion presented here allows us to conclude that the phase transition from limited to unlimited energy growth in a dissipative oval billiard is analogous to a second-order phase transition.

References

1. Almeida AMO (1995) Sistemas hamiltonianos: caos e quantização. Campinas, Editora da Unicamp
2. Aguilar-Sanchez R, Leonel ED, Méndez-Bermúdez JA (2013) Dynamical properties of a dissipative discontinuous map: A scaling investigation. Phys Lett A 377:3216
3. Bai-lin H (1990) Chaos II. World Scientific Publishing Co. Pte. Ltd., Singapore
4. Balakrishnan V (2008) Elements of nonequilibrium statistical mechanics. Ane Books India, New Delhi
5. Berry MV (1981) Regularity and chaos in classical mechanics, illustrated by three deformations of a circular billiard. Eur J Phys 2(2):91–102
6. Birkhoff G (1927) Dynamical systems. American Mathematical Society, Providence, RI, USA
7. Bonatto C, Garreau JC, Gallas JAC (2005) Self similarities in the frequency-amplitude space of a loss-modulated CO2 laser. Phys Rev Lett 95:143905
8. Botari T, Leonel ED (2013) A one-dimensional Fermi accelerator model with moving wall described by a nonlinear van der Pol oscillator. Phys Rev E 87:012904
9. Bunimovich LA (1974) On the ergodic properties of certain billiards. Funct Anal Appl 8:73–74
10. Bunimovich LA, Sinai YG (1980) Statistical properties of Lorentz gas with periodic configuration of scatterers. Commun Math Phys 78(4):479–497
11. Butkov E (1988) Física Matemática. Livraria Técnica Científica, Guanabara
12. Carvalho RE, Souza FC, Leonel ED (2006) Fermi acceleration on the annular billiard. Phys Rev E 73(6):66229
13. Carvalho RE, Souza FC, Leonel ED (2006) Fermi acceleration on the annular billiard: A simplfied version. J Phys A: Math Gen 39:3561–3573
14. Chernov N, Markarian R (2006) Chaotic billiards, vol 127. American Mathematical Society, Providence, RI, USA
15. Chirikov BV (1979) A universal instability of many-dimensional oscillator systems. Phys Rep 52(5):263–379
16. Churchill RV, Brown JW (1984) Complex variables with applications. McGraw-Hill Book Company, Singapore
17. Costa DR, Carvalho RE (2018) Dynamics of a light beam suffering the influence of a dispersing mechanism with tunable refraction index. Phys Rev E 98:022224
18. Costa D, Dettmann C, Leonel ED (2011) Escape of particles in a time-dependent potential well. Phys Rev E 83:066211

19. Costa D, Livorati ALP, Leonel ED (2012) Critical exponents and scaling properties for the chaotic dynamics of a particle in a time-dependent potential barrier. Int J Bifurcat Chaos Appl Sci Eng 22:1250250

20. Costa DR, Silva MR, Oliveira JA, Leonel ED (2012) Scaling dynamics for a particle in a time-dependent potential well. Phys A 391:3607

21. Dettmann CP, Leonel ED (2012) Escape and transport for an open bouncer: Stretched exponential decays. Phys D 241:403

22. Devaney RL (1992) A first course in chaotic dynamical systems: Theory and experiment. Addison-Wesley Publishing Company Inc, Massachusetts

23. Devaney RL (2003) An introduction to chaotic dynamical systems. Westview Press, Cambridge

24. Díaz-Iturry G, Livorati ALP, Leonel ED (2016) Statistical investigation and thermal properties for a 1-D impact system with dissipation. Phys Lett A 380:1830

25. Eckmann J-P, Ruelle D (1985) Ergodic theory of chaos and strange attractors. Rev Mod Phys 57:617

26. Everson RM (1986) Chaotic dynamics of a bouncing ball. Phys D 19(3):355–383

27. Faria NB, Tavares DS, Paula WCS, Leonel ED, Ladeira DG (2016) Transport of chaotic trajectories from regions distant from or near to structures of regular motion of the Fermi-Ulam model. Phys Rev E 94:042208

28. Fatunla SO (1988) Numerical methods for initial value problems in ordinary differential equations. Academic Press Inc, San Diego

29. Galia MVC, Oliveira DFM, Leonel ED (2016) Evolution to the equilibrium in a dissipative and time dependent billiard. Phys A 465:66

30. Gleick J (1989) Chaos. Sphere Books Ltd, London

31. Grasman J (1987) Asymptotic methods for relaxation oscillations and applications. Springer-Verlag, New York

32. Grebogi C, Ott E, Pelikan S, Yorke JA (1984) Strange attractors that are not chaotic. Phys D 13:261

33. Grebogi C, Ott E, Yorke JA (1986) Critical exponent of chaotic transients in nonlinear dynamical systems. Phys Rev Lett 57(11):1284–1287

34. Grebogi C, Ott E, Romeiras F, Yorke JA (1987) Critical exponents for crisis-induced intermittency. Phys Rev A 36:5365

35. Guckenheimer J, Holmes P (1983) Nonlinear oscillations, dynamical systems, and bifurcations of vector fields. Springer-Verlag, New York

36. Hansen M, Ciro D, Caldas IL, Leonel ED (2018) Explaining a changeover from normal do super diffusion in time-dependent billiards. Europhys Lett 121(1):60003

37. Hansen M, Ciro D, Caldas IL, Leonel ED (2019) Dynamical thermalization in time-dependent billiards. Chaos 29:103122

38. Hansen M, Costa DR, Caldas IL, Leonel ED (2018) Statistical properties for an open oval billiard: An investigation of the escaping basins. Chaos Solitons Fractals 106:355

39. Hermes JDV, Graciano FH, Leonel ED (2020) Universal behavior of the convergence to the stationary state for a tangent bifurcation in the logistic map. Discont Nonlinearity Complexity 9:63

40. Hilborn RC (1986) Chaos and nonlinear dynamics. Oxford University Press, New York

41. Holmes PJ (1982) The dynamics of repeated impacts with asinusoidally vibrating table. J Sound Vibr 84(2):173–189

42. Horstmann AC, Holokx AA, Manchein C (2017) The effect of temperature on generic stable periodic structures in the parameter space of dissipative relativistic standard map. Eur Phys J B 90(5):96

43. Howard JE, Humpherys J (1995) Nonmonotonic twist maps. Phys D 80(3):256–276

44. Iserles A (2009) A first course in the numerical analysis of differential equations. Cambridge University Press, Cambridge

45. Jousseph CA, Abdulack SA, Manchein C, Beims MW (2018) Hierarchical collapse of regular islands via dissipation. J Phys A: Math Theoret 51:105101

46. Karlis AK, Papachristou PK, Diakonos FK, Constantoudis V, Schmelcher P (2006) Hyperacceleration in a stochastic fermi-ulam model. Phys Rev Lett 97:194102
47. Krylov NS (1979) Works on the foundations of statistical physics. Princeton University Press, Princeton, NJ, USA
48. Kuwana CM, Oliveira JA, Leonel ED (2013) A family of dissipative two-dimensional mappings: Chaotic, regular and steady state dynamics investigation. Phys A 395:458
49. Ladeira DG, Leonel ED (2007) Dynamical properties of a dissipative hybrid Fermi-Ulambouncer model. Chaos 17(1):013119
50. Ladeira DG, Leonel ED (2010) Competition between suppression and production of Fermi acceleration. Phys Rev E 81:036216
51. Lenz F, Diakonos FK, Schmelcher P (2008) Tunable fermi acceleration in the driven elliptical billiard. Phys Rev Lett 100(1):014103(1)–014103(4)
52. Leonel ED (2015) Fundamentos da Física Estatística. São Paulo, Blucher
53. Leonel ED (2022) Scaling laws in dynamical systems. Higher Educational Press, Beijing
54. Leonel ED (2019) Invariância de Escala em Sistemas Dinâmicos Não Lineares. São Paulo, Blucher
55. Leonel ED (2007) Breaking down the Fermi acceleration with inelastic collisions. J Phys A: Math Theor 40:F1077–F1083
56. Leonel ED (2007) Corrugated waveguide under scaling investigation. Phys Rev Lett 98:114102(1)–114102(4)
57. Leonel ED (2009) Phase transition in dynamical systems: defining classes of universality for two-dimensional hamiltonian mappings via critical exponents. Math Prob Eng 2009:1–22
58. Leonel ED (2016) Defining universality classes for three different local bifurcations. Commun Nonlinear Sci Numer Simul 39:520
59. Leonel ED, Bunimovich LA (2010) Suppressing Fermi acceleration in a driven elliptical billiard. Phys Rev Lett 104:224101
60. Leonel ED, Bunimovich L (2010) Suppressing Fermi acceleration in two-dimensional driven billiards. Phys Rev E 82:016202
61. Leonel ED, Carvalho RE (2007) A family of crisis in a dissipative Fermi accelerator model. Phys Lett A 364:475
62. Leonel ED, Galia MVC, Barreiro LA, Oliveirs DFM (2016) Thermodynamics of a time-dependent and dissipative oval billiard: A heat transfer and billiard approach. Phys Rev E 94:062211
63. Leonel ED, Kuwana CM (2017) An investigation of chaotic diffusion in a family of hamiltonian mappings whose angles diverge in the limit of vanishingly action. J Stat Phys 170:69
64. Leonel ED, Kuwana CM, Yoshida M, Oliveira JA (2020) Application of the diffusion equation to prove scaling invariance on the transition from limited to unlimited diffusion. EPL (Europhys Lett) 131:10004
65. Leonel ED, Kuwana CM, Yoshida M, Oliveira JA (2020) Chaotic diffusion for particles moving in a time dependent potential well. Phys Lett A 384:126737
66. Leonel ED, Livorati ALP (2008) Describing Fermi acceleration with a scaling approach: the bouncer model revisited. Phys A: Stat Mech Appl 387(5–6):1155–1160
67. Leonel ED, Livorati ALP (2015) Thermodynamics of a bouncer model: A simplified one-dimensional gas. Commun Nonlinear Sci Numer Simul 20:159–173
68. Leonel ED, Livorati ALP, Céspedes AM (2014) A theoretical characterization of scaling properties in a bouncing ball system. Phys A: Stat Mech Appl 404:279–284
69. Leonel ED, Marinho EP (2009) Fermi acceleration with memory-dependent excitation. Phys A 388:4927
70. Leonel ED, McClintock PVE (2005) A hybrid Fermi-Ulam-bouncer model. J Phys A: Math Gen 3(4):823
71. Leonel ED, McClintock PVE (2005) Scaling properties for a classical particle in a time-dependent potential well. Chaos 15:033701
72. Leonel ED, McClintock PVE (2004) Dynamical properties of a particle in a time-dependent double-well potential. J Phys A: Math Gen 37:8949

73. Leonel ED, McClintock PVE (2004) Chaotic properties of a time-modulated barrier. Phys Rev E 70:16214
74. Leonel ED, McClintock PVE (2005) A crisis in the dissipative Fermi accelerator model. J Phys A: Math Gen 38:L425
75. Leonel ED, McClintock PVE (2006) Effect of a frictional force on the Fermi-Ulam model. J Phy A: Math Gen 39:11399
76. Leonel ED, McClintock PVE (2006) Dissipative area-preserving one-dimensional Fermi accelerator model. Phys Rev E 73:66223
77. Leonel ED, McClintock PVE, Silva JKL (2004) The Fermi-Ulam accelerator model under scaling analysis. Phys Rev Lett 93:14101
78. Leonel ED, Oliveira DFM, Loskutov A (2009) Fermi acceleration and scaling properties of a time dependent oval billiard. Chaos 19:033142
79. Leonel ED, Oliveira JA, Saif F (2011) Critical exponents for a transition from integrability to non-integrability via localization of invariant tori in the Hamiltonian system. J Phys A: Math Theoret 44:302001(1)–302001(7)
80. Leonel ED, Penalva J, Teixeira RMN, Costa-Filho RN, Silva MR, Oliveira JA (2015) A dynamical phase transition for a family of Hamiltonian mappings: A phenomenological investigation to obtain the critical exponents. Phys Lett A 1808
81. Leonel ED, Silva MR (2008) A bouncing ball model with two nonlinearities: a prototype for Fermi acceleration. J Phys A: Math Theoret 41:015104(1)–015104(13)
82. Leonel ED, Silva JKL (2003) Dynamical properties of a particle in a classical time-dependent potential well. Phys A 323:181
83. Leonel ED, Silva JKL, Kamphorst SO (2004) On the dynamical properties of a Fermi accelerator model. Phys A 331:435
84. Leonel ED, Yoshida M, Oliveira JA (2020) Characterization of a continuous phase transition in a chaotic system. EPL (Europhys Lett) 131:20002
85. Lieberman MA, Lichtenberg AJ (1972) Stochastic and adiabatic behavior of particles accelerated by periodic forces. Phys Rev A 5:1852
86. Lindner JF, Kohar V, Kia B, Hippke M, Learned JG, Ditto WL (2015) Strange nochaotic stars. Phys Rev Lett 114:054101
87. Livorati ALP et al (2012) Stickiness in a bouncer model: A slowing mechanism for Fermi acceleration. Phys Rev E 86(3):9
88. Livorati ALP, Dettmann CP, Caldas IL, Leonel ED (2015) On the statistical and transport properties of a non-dissipative Fermi-Ulam model. Chaos 25:103107
89. Livorati ALP, Kroetz T, Dettmann CP, Caldas IL, Leonel ED (2012) Stickiness in a bouncer model: A slowing mechanism for Fermi acceleration. Phys Rev E 86:036203
90. Livorati ALP, Ladeira DG, Leonel ED (2008) Scaling investigation of Fermi acceleration on a dissipative bouncer model. Phys Rev E 78:056205
91. Livorati ALP, Loskutov A, Leonel ED (2011) A family of stadium-like billiards with parabolic boundaries under scaling analysis. J Phys A: Math Theoret 44:175102
92. Livorati ALP, Oliveira JA, Ladeira DG, Leonel ED (2014) Time-dependent properties in two-dimensional and Hamiltonian mappings. Eur Phys J. 223:2953
93. Livorati ALP, Palmero MS, Dettmann CP, Caldas IL, Leonel ED (2014) Separation of particles leading either to decay or unlimited growth of energy in a driven stadium-like billiard. J Phys A: Math Theoret 47:365101
94. Lorenz EN (1963) Deterministic nonperiodic flow. J Atmos Sci 20:130
95. Loskutov A, Ryabov AB, Akinshin LG (2000) Properties of some chaotic billiards with time-dependent boundaries. J Phys A: Math Gen 33:7973–7986
96. Luna-Acosta GA (1990) Regular and chaotic dynamics of the damped Fermi accelerator. Phys Rev A: At Mol Opt Phys 42(12):7155–7162
97. Manchein C, Beims MW (2013) Conservative generalized bifurcation diagrams. Phys Lett A 377:789
98. Manchein C, Beims MW (2009) Dissipation effects in the ratchetlike Fermi acceleration. Math Probl Eng 2009:513023

99. Manneville P (2004) Instabilities, chaos and turbulence: An introduction to nonlinear dynamics and complex systems. Imperial College Press, London
100. Méndez-Bermúdez JA, Oliveira JA, Aguilar-Sánchez R, Leonel ED (2015) Scaling properties for a family of discontinuous mappings. Phys A 436:943
101. Méndez-Bermúdez JA, Oliveira JA, Leonel ED (2016) Analytical description of critical dynamics for two-dimensional dissipative nonlinear maps. Phys Lett A 380:1959
102. Mendonca HMJ, Leonel ED, Oliveira JA (2016) An investigation of the convergence to the stationary state in the Hassell mapping. Phys A 466:537
103. Monteiro LHA (2011) Sistemas dinâmicos. São Paulo, Livraria da Física
104. Mugnaine M, Mathias AC, Santos MS, Batista AM, Szezech JRJD, Viana RL (2018) Dynamical characterization of transport barriers in nontwist Hamiltonian systems. Phys Rev E 97(1):012214
105. Oliveira JA, Bizão RA, Leonel ED (2010) Finding critical exponents for two-dimensional Hamiltonian maps. Phys Rev E 81:046212
106. Oliveira JA, Dettmann CP, Costa DR, Leonel ED (2013) Scaling invariance of the diffusion coefficient in a family of two-dimensional Hamiltonian mappings. Phys Rev E 87:062904
107. Oliveira JA, Leonel ED (2012) Dissipation and its consequences in the scaling exponents for a family of two-dimensional mappings. J Phys A: Math Theoret 45:165101
108. Oliveira DFM, Leonel ED (2012) Dynamical properties for the problem of a particle in a electric field of wave packet: Low velocity and relativistic approach. Phys Lett A 376:3630
109. Oliveira DFM, Leonel ED (2013) Some dynamical properties of a classical dissipative bouncing ball model with two nonlinearities. Phys A 392:1762
110. Oliveira DFM, Leonel ED (2012) In-flight and collisional dissipation as a mechanism to suppress Fermi acceleration in a breathing Lorentz gas. Chaos 22:026123
111. Oliveira DFM, Leonel ED (2008) The feigenbaum's δ for a high dissipative bouncing ball model. Braz J Phys 38(1):62–64
112. Oliveira DFM, Leonel ED (2010) Suppressing Fermi acceleration in a two-dimensional nonintegrable time-dependent oval-shaped billiard with inelastic collisions. Phys A 389:1009
113. Oliveira JA, Leonel ED (2013) A rescaling of the phase space for Hamiltonian map: Applications on the Kepler map and mappings with diverging angles in the limit of vanishing action. Appl Math Comput 221:32–39
114. Oliveira JA, Leonel ED (2011) Locating invariant tori for a family of two-dimensional Hamiltonian mappings. Phys A 390:3727–3731
115. Oliveira DFM, Leonel ED, Robnik M (2011) Boundary crisis and transient in a dissipative relativistic standard map. Phys Lett A 375:3365
116. Oliveira JA, Papesso ER, Leonel ED (2013) Relaxation to fixed points in the logistic and cubic maps: Analytical and numerical investigation. Entropy 15:4310
117. Oliveira DFM, Robnik M, Leonel ED (2012) Statistical properties of a dissipative kicked system: Critical exponents and scaling invariance. Phys Lett A 376(5):723–728
118. Oliveira DFM, Robnik M, Leonel ED (2011) Dynamical properties of a particle in a wave packet: Scaling invariance and boundary crisis. Chaos Solitons Fractals 44:883
119. Oliveira DFM, Vollmer J, Leonel ED (2011) Fermi acceleration and its suppression in a time-dependent Lorentz gas. Phys D 240:389
120. Oliveira DFM, Silva MR, Leonel ED (2015) A symmetry break in energy distribution and a biased random walk behavior causing unlimited diffusion in a two dimensional mapping. Phys A 436:909
121. Ott E (2002) Chaos in dynamical systems. Cambridge University Press, Cambridge
122. Palmero MS, Díaz-Iturry G, McClintock PVE, Leonel ED (2020) Diffusion phenomena in a mixed phase space. Chaos 30:013108
123. Palmero MS, Livorati ALP, Caldas IL, Leonel ED (2018) Ensemble separation and stickiness influence in a driven stadium-like billiard: A Lyapunov exponents analysis. Commun Nonlinear Sci Numer Simul 65:248
124. Pereira T, Baptista MS, Reys MB, Caldas IL, Sartorelli JC, Kurths J (2006) Global bifurcations destroying the experimental torus T2. Phys Rev E 73:01720(1)–01720(4)

125. Press WH, Flannery BP, Teukolsky SA, Vetterling WT (1992) Numerical recipes in fortran 77: The art of scientific computing, 2nd edn. Cambridge University Press

126. Portela JS, Caldas IL, Viana RL, Morrison PJ (2007) Diffusive transport through a nontwist barrier in tokamaks. Int J Bifurcat Chaos Appl Sci Eng 17:1589–1598

127. Pustylnikov LD (1978) Transactions of the moscow mathematical society. Providence 2:1

128. Rabelo AF, Leonel ED (2008) Finding invariant tori in the problem of a periodically corrugated waveguide. Braz J Phys 38:54

129. Rech PC (2008) Naimark-sacker bifurcations in a delay quartic map. Chaos Solitons Fractals 37:387

130. Romeiras FJ, Ott E (1987) Strange nonchaotic attractors of the damped pendulum with quasiperiodic forcing. Phys Rev A 35:4404

131. Saltzman B (1962) Finite amplitude free convenction as an initial value problem-I. J Atmos Sci 19:329

132. Santos MS, Mugnaine M, Szezech JD, Batista AM, Caldas IL, Viana RL (2019) Using rotation number to detect sticky orbits in Hamiltonian systems. Chaos 29:043125

133. Sethna JP (2006) Entropy, order parameters, and complexity. Oxford University Press, Oxford

134. Silva MR, Costa DR, Leonel ED (2012) Characterization of multiple reflections and phase space properties for a periodically corrugated waveguide. J Phys A: Math Theoret 45:265101

135. Silva VB, Leonel ED (2018) Evolution towards the steady state in a hopf bifurcation: A scaling investigation. Discontinuity Nonlinearity Complexity 7:67

136. Silva JKL, Leonel ED, McClintock PVE, Kamphorst SO (2006) Scaling properties of the Fermi-Ulam accelerator model. Braz J Phys 36(3a):700–707

137. Silva RM, Manchein C, Beims MW, Altmann EG (2015) Characterizing weak chaos using time series of Lyapunov exponents. Phys Rev E 91:062907

138. Sinai YG (1970) Dynamical systems with elastic reflections. Russ Math Surv 25(2):137–189

139. Souza FA, Simões LEA, Silva MR, Leonel ED (2009) Can drag force suppress fermi acceleration in a bouncer model? Math Probl Eng 2009:1

140. Strogatz SH (2015) Nonlinear dynamics and chaos: With applications to physics, biology, chemistry and engineering. Westview Press, Bolder

141. Tabor M (1989) Chaos and integrability in nonlinear dynamics: An introduction. John Wiley & Sons, New York

142. Tavares DF, Araujo AD, Leonel ED, Costa-Filho RN (2013) Dynamical properties for a mixed Fermi accelerator model. Phys A 392:4231–4241

143. Tavares DF, Leonel ED (2008) A simplified fermi accelerator model under quadratic frictional force. Braz J Phys 38(1):58–61

144. Tavares DF, Leonel ED, Costa-Filho RN (2012) Non-uniform drag force on the Fermi accelerator model. Phys A 391:5366–5374

145. Teixeira RMN, Rando DS, Geraldo FC, Costa-Filho RN, Oliveira JA, Leonel ED (2015) Convergence towards asymptotic state in 1-D mappings: A scaling investigation. Phys Lett A 379:1246

146. Wiggers V, Rech PC (2017) Multistability and organization of periodicity in a Van der Pol-Duffing oscillator. Chaos Solitons Fractals 103:632–637

147. Wiggers V, Rech PC (2017) Chaos, periodicity, and quasiperiodicity in a radio-physical oscillator. Int J Bifurcat Chaos 27(7):1730023

148. Wiggins S (1990) Introduction to applied nonlinear dynamical systems and chaos. Springer-Verlag, New York

scan this code for color figures

NONLINEAR PHYSICAL SCIENCE

(Series Editors: Albert C.J. Luo, Dimitri Volchenkov)

ISBN 978-7-04-061663-7	57 Dynamical Phase Transitions in Chaotic Systems (2024) by Edson Denis Leonel
ISBN 978-7-04-060211-1	56 Two-dimensional Quadratic Nonlinear Systems: Bivariate Vector Fields (2023) by Albert C. J. Luo
ISBN 978-7-04-060495-5	55 Two-Dimensional Quadratic Nonlinear Systems: Univariate Vector Fields (2023) by Albert C. J. Luo
ISBN 978-7-04-059634-2	54 Principles of Innovative Design Thinking (2023) by Wenjuan Li, Zhenghe Song, C. Steve Such
ISBN 978-7-04-058735-7	53 Symmetries and Applications of Differential Equations (2022) by Albert C. J. Luo, Rafail K. Gazizov(Editors)
ISBN 978-7-04-060822-9	52 Mathematical Topics on Modelling Complex Systems （2023） by J.A. Tenreiro Machado，Dimitri Volchenkov(Editors)
ISBN 978-7-04-057152-3	51 The Many Facets of Complexity Science (2022) by Dimitri Volchenkov (Editor)
ISBN 978-7-04-057212-4	50 Scaling Laws in Dynamical Systems (2022) by Edson Denis Leonel
ISBN 978-7-04-055911-8	49 Nonlinear Dynamics, Chaos, and Complexity (2021) by Dimitri Volchenkov (Editor)
ISBN 978-7-04-055802-9	48 Slowly Varying Oscillations and Waves (2021) by Lev Ostrovsky
ISBN 978-7-04-058436-3	47 不连续动力系统 (2022) 罗朝俊著，闵富红、李欣业译
ISBN 978-7-04-054753-5	46 连续动力系统 (2021) 罗朝俊著，王跃方、黄金、李欣业译
ISBN 978-7-04-055783-1	45 Bifurcation Dynamics in Polynomial Discrete Systems (2021) by Albert C. J. Luo
ISBN 978-7-04-055228-7	44 Bifurcation and Stability in Nonlinear Discrete Systems (2020) by Albert C. J. Luo

ISBN 978-7-04-050615-0	43 Theory of Hybrid Systems: Deterministic and Stochastic (2019) by Mohamad S. Alwan, Xinzhi Liu
ISBN 978-7-04-050235-0	42 Rigid Body Dynamics: Hamiltonian Methods, Integrability, Chaos (2018) by A. V. Borisov, I. S. Mamaev
ISBN 978-7-04-048458-8	41 Galloping Instability to Chaos of Cables (2018) by Albert C. J. Luo, Bo Yu
ISBN 978-7-04-048004-7	40 Resonance and Bifurcation to Chaos in Pendulum (2017) by Albert C. J. Luo
ISBN 978-7-04-047940-9	39 Grammar of Complexity: From Mathematics to a Sustainable World (2017) by Dimitri Volchenkov
ISBN 978-7-04-047809-9	38 Type-2 Fuzzy Logic: Uncertain Systems' Modeling and Control (2017) by Rómulo Martins Antão, Alexandre Mota, R. Escadas Martins, J. Tenreiro Machado
ISBN 978-7-04-047450-3	37 Bifurcation in Autonomous and Nonautonomous Differential Equations with Discontinuities (2017) by Marat Akhmet, Ardak Kashkynbayev
ISBN 978-7-04-043231-2	36 离散和切换动力系统（中文版）(2015) 罗朝俊
ISBN 978-7-04-043102-5	35 Replication of Chaos in Neural Networks, Economics and Physics (2015) by Marat Akhmet, Mehmet Onur Fen
ISBN 978-7-04-042835-3	34 Discretization and Implicit Mapping Dynamics (2015) by Albert C.J.Luo
ISBN 978-7-04-042385-3	33 Tensors and Riemannian Geometry with Applications to Differential Equations (2015) by Nail Ibragimov
ISBN 978-7-04-042131-6	32 Introduction to Nonlinear Oscillations (2015) by Vladimir I. Nekorkin
ISBN 978-7-04-038891-6	31 Keller-Box Method and Its Application (2014) by K. Vajravelu, K.V. Prasad
ISBN 978-7-04-039179-4	30 Chaotic Signal Processing (2014) by Henry Leung
ISBN 978-7-04-037357-8	29 Advances in Analysis and Control of Time-Delayed Dynamical Systems (2013) by Jianqiao Sun, Qian Ding (Editors)

ISBN 978-7-04-036944-1	28 Lectures on the Theory of Group Properties of Differential Equations (2013) by L.V. Ovsyannikov (Author), Nail Ibragimov (Editor)
ISBN 978-7-04-036741-6	27 Transformation Groups and Lie Algebras (2013) by Nail Ibragimov
ISBN 978-7-04-030734-4	26 Fractional Derivatives for Physicists and Engineers Volume II. Applications (2013) by Vladimir V. Uchaikin
ISBN 978-7-04-032235-4	25 Fractional Derivatives for Physicists and Engineers Volume I. Background and Theory (2013) by Vladimir V. Uchaikin
ISBN 978-7-04-035449-2	24 Nonlinear Flow Phenomena and Homotopy Analysis: Fluid Flow and Heat Transfer (2012) by Kuppalapalle Vajravelu, Robert A.Van Gorder
ISBN 978-7-04-034819-4	23 Continuous Dynamical Systems (2012) by Albert C.J. Luo
ISBN 978-7-04-034821-7	22 Discrete and Switching Dynamical Systems (2012) by Albert C.J. Luo
ISBN 978-7-04-032279-8	21 Pseudo chaotic Kicked Oscillators: Renormalization, Symbolic Dynamics, and Transport (2012) by J.H. Lowenstein
ISBN 978-7-04-032298-9	20 Homotopy Analysis Method in Nonlinear Differential Equations (2011) by Shijun Liao
ISBN 978-7-04-031964-4	19 Hyperbolic Chaos: A Physicist's View (2011) by Sergey P. Kuznetsov
ISBN 978-7-04-032186-9	18 非线性变形体动力学 （中文版）(2011) 罗朝俊著，郭羽、黄健哲、闵富红译
ISBN 978-7-04-031954-5	17 Applications of Lie Group Analysis in Geophysical Fluid Dynamics (2011) by Ranis Ibragimov, Nail Ibragimov
ISBN 978-7-04-031957-6	16 Discontinuous Dynamical Systems (2011) by Albert C.J. Luo
ISBN 978-7-04-031694-0	15 Linear and Nonlinear Integral Equations: Methods and Applications (2011) by Abdul-Majid Wazwaz
ISBN 978-7-04-029710-2	14 Complex Systems: Fractionality, Time-delay and Synchronization (2011) by Albert C.J. Luo , Jianqiao Sun (Editors)

ISBN 978-7-04-031534-9	13 Fractional-order Nonlinear Systems: Modeling, Analysis and Simulation (2011) by Ivo Petráš
ISBN 978-7-04-031533-2	12 Bifurcation and Chaos in Discontinuous and Continuous Systems (2011) by Michal Fečkan
ISBN 978-7-04-031695-7	11 Waves and Structures in Nonlinear Nondispersive Media: General Theory and Applications to Nonlinear Acoustics (2011) by S.N. Gurbatov, O.V. Rudenko, A.I. Saichev
ISBN 978-7-04-032187-6	10动态域上的不连续动力系统 （中文版）(2011) 罗朝俊著，闵富红，黄健哲，郭羽译
ISBN 978-7-04-029474-3	9 Self-organization and Pattern-formation in Neuronal Systems Under Conditions of Variable Gravity (2011) by Meike Wiedemann, Florian P.M. Kohn, Harald Rosner, Wolfgang R.L. Hanke
ISBN 978-7-04-029473-6	8 Fractional Dynamics (2010) by Vasily E. Tarasov
ISBN 978-7-04-029187-2	7 Hamiltonian Chaos beyond the KAM Theory (2010) by Albert C.J. Luo, Valentin Afraimovich (Editors)
ISBN 978-7-04-029188-9	6 Long-range Interactions, Stochasticity and Fractional Dynamics (2010) by Albert C.J. Luo, Valentin Afraimovich (Editors)
ISBN 978-7-04-028882-7	5 Nonlinear Deformable-body Dynamics (2010) by Albert C.J. Luo
ISBN 978-7-04-018292-7	4 Mathematical Theory of Dispersion-Managed Optical Solitons (2010) by A. Biswas, D. Milovic, E. Matthew
ISBN 978-7-04-025480-8	3 Partial Differential Equations and Solitary Waves Theory (2009) by Abdul-Majid Wazwaz
ISBN 978-7-04-025759-5	2 Discontinuous Dynamical Systems on Time-varying Domains (2009) by Albert C.J. Luo
ISBN 978-7-04-025159-3	1 Approximate and Renormgroup Symmetries (2009) by Nail H. Ibargimov

郑重声明

高等教育出版社依法对本书享有专有出版权。任何未经许可的复制、销售行为均违反《中华人民共和国著作权法》，其行为人将承担相应的民事责任和行政责任；构成犯罪的，将被依法追究刑事责任。为了维护市场秩序，保护读者的合法权益，避免读者误用盗版书造成不良后果，我社将配合行政执法部门和司法机关对违法犯罪的单位和个人进行严厉打击。社会各界人士如发现上述侵权行为，希望及时举报，我社将奖励举报有功人员。

反盗版举报电话	(010) 58581999　58582371
反盗版举报邮箱	dd@hep.com.cn
通信地址	北京市西城区德外大街 4 号
	高等教育出版社法律事务部
邮政编码	100120